AF587519

Reliable and Robust Optimal Design of Sustainable Energy Systems

Zuverlässige und robuste optimale Auslegung von nachhaltigen Energiesystemen

Von der Fakultät für Maschinenwesen der Rheinisch-Westfälischen Technischen Hochschule Aachen zur Erlangung des akademischen Grades einer Doktorin der Ingenieurwissenschaften genehmigte Dissertation

vorgelegt von

Dinah Elena Hollermann (geb. Majewski)

Berichter: Univ.-Prof. Dr.-Ing. André Bardow
Univ.-Prof. Dr.-Ing. Veith Hagenmeyer

Tag der mündlichen Prüfung: 17. Februar 2020

Diese Dissertation ist auf den Internetseiten der Universitätsbibliothek online verfügbar.

Aachener Beiträge zur Technischen Thermodynamik Band 26

Dinah Elena Hollermann
Reliable and Robust Optimal Design of Sustainable Energy Systems

Zuverlässige und robuste optimale Auslegung von nachhaltigen Energiesystemen

ISBN: 978 3 95886-346-0

Bibliografische Information der Deutschen Bibliothek
Die Deutsche Bibliothek verzeichnet diese Publikation in der Deutschen Nationalbibliografie; detaillierte bibliografische Daten sind im Internet über http://dnb.ddb.de abrufbar.

Herstellung & Vertrieb:

1. Auflage 2020
© Wissenschaftsverlag Mainz GmbH - Aachen
Süsterfeldstr. 83, 52072 Aachen
Tel. 0241 / 87 34 34 00
www.Verlag-Mainz.de

ISSN: 2198-4832

Satz: nach Druckvorlage des Autors
Umschlaggestaltung: Druckerei Mainz

printed in Germany
D82 (Diss. RWTH Aachen University, 2020)

True optimization is the revolutionary contribution of modern research to decision processes.

George B. Dantzig (1904 - 2005)

Vorwort

Die vorliegende Arbeit habe ich im Rahmen meiner Tätigkeit als wissenschaftliche Mitarbeiterin am Lehrstuhl für Technische Thermodynamik der RWTH Aachen University verfasst. Besonderer Dank gilt meinem Doktorvater Prof. André Bardow. Danke für die konstruktiven fachlichen Diskussionen und die Wertschätzung meiner Arbeit sowie für das mir entgegengebrachte Vertrauen und die Möglichkeit das Optimierungs-Team leiten zu dürfen. Du hast maßgeblich zu meiner persönlichen und fachlichen Entwicklung beigetragen. Schön, dass das alles auch mit Kind möglich war und ich sehr viel Flexibilität genießen durfte. Weiterer Dank gebührt Prof. Veith Hagenmeyer für die gewissenhafte Übernahme des Koreferats. Ich danke Prof. Dirk Müller für den lockeren und trotzdem professionellen Prüfungsvorsitz.

Sowohl Prof. Veit Hagenmeyer als auch Prof. Dirk Müller danke ich zudem für die gute gemeinsame Arbeit in der Helmholtz-Initiative *Energie System 2050 – Ein Beitrag des Forschungsbereichs Energie*. Darüber hinaus bedanke ich mich bei Timm Faulwasser, Thomas Schütz, Dominique Sauer für die konstruktive Zusammenarbeit im Rahmen des Forschungsprojektes und besonders bei Susanne Sass für das Voranbringen unserer Forschung. Marc Goerigk danke ich für das gemeinsame Paper, das hoffentlich noch mehr anwendungsorientierte Mathematik in die Methoden der Ingenieurswissenschaften bringt.

Ich habe mich am Lehrstuhl für Technische Thermodynamik sehr wohl gefühlt. Eine lockere, gemeinschaftliche und trotzdem sehr produktive Arbeitsatmosphäre haben mir Spaß gemacht und mich motiviert. Hervorzuheben sind meine Gruppenleiter/-in Philip Voll, Matthias Lampe und Maike Hennen, die sich viel Zeit für Diskussionen über meine Forschung genommen haben. Es war schön, am Lehrstuhl nicht nur wissenschaftlich zusammen zu arbeiten, sondern darüber hinaus noch viele gemeinsame private Aktionen zu genießen. Danke dafür! Besonders bedanke ich mich bei meinen Bürokollegen/-innen Maike Hennen, Dominik Tillmanns, Christiane Reinert, Fabian Mayer, Dörthe Hoffrogge und meinem Schreibtischkollegen Nils Baumgärtner, die sich meine Klagen anhören durften, wenn es einmal nicht so gut lief, mit denen ich aber auch eine wundervolle Zeit hatte.

Darüber hinaus danke ich allen Studenten/-innen, die zum Gelingen dieser Arbeit beigetragen haben, besonders Marco Wirtz, Malte Jensen, Maren Ihlemann, Fabian Mayer und Dörthe Hoffrogge.

Danke an die Akrobatik-Gruppe in Aachen, meine Nachbarn Stephan und Hannah sowie meine Eltern-Peergroup für alles was nichts mit Arbeit zu tun hat.

Der größte Dank gilt meiner Familie: Ich danke meinen Eltern, dass Sie mich immer in meinen Vorhaben unterstützt und mir Mut gemacht haben. Danke, dass ihr immer für mich da seid! Meinem Bruder Dennis danke ich dafür, dass er der Große ist und mir einige Wege gebahnt hat. Henning, ich danke dir für das Rückenfreihalten, dein Verständnis, die Ablenkung, die unkonventionellen Urlaube, den Spaß, das Vertrauen, die Offenheit und für Timo, der mich in den letzten Jahren viel zum Lachen gebracht hat und mir immer wieder zeigt, was wirklich wichtig ist. Ihr macht mich glücklich!

Kassel, März 2020

Kurzfassung

Die Auslegung von Energiesystemen ist eine komplexe Aufgabe, da viele Faktoren den Entscheidungsprozess beeinflussen. Mathematische Optimierung ist ein hervorragender Ansatz, um diese komplexe Aufgabe zu lösen und ein optimales Energiesystem zu identifizieren. Die Optimierung von Energiesystemen ist jedoch intrinsisch unsicher, da sowohl die Verfügbarkeit der Energieversorgungsanlagen als auch Optimierungsparameter, wie beispielsweise Energiebedarfe, unsicher sind. Die Vernachlässigung von Unsicherheiten kann zu einem Defizit in der Energieversorgung führen, das zu unerwartet hohen Kosten und Umweltschäden führen kann. Aus diesem Grund präsentiert diese Dissertation Methoden, die Unsicherheiten bereits während der Optimierung berücksichtigen. Neben Unsicherheiten ist Nachhaltigkeit ein wichtiges Thema in der Energiesystemoptimierung. Nachhaltigkeit umfasst dabei die Aspekte Ökonomie, Ökologie und Soziales. Die Anwendung von multikriterieller Optimierung ist daher hervorragend geeignet, um die Aspekte der Nachhaltigkeit zu berücksichtigen. Multikriterielle Optimierung führt jedoch in der Regel zu einer Vielzahl von Designvarianten, von denen jedoch ein einziges Design ausgewählt werden muss.

Die vorliegende Arbeit stellt eine Methode zur Auswahl des besten zuverlässigen und robusten Designs von nachhaltigen Energiesystemen vor. Die Methode berücksichtigt sowohl Unsicherheiten in der Verfügbarkeit der Energieversorgungsanlagen als auch in den Eingangsparametern der Optimierung. Daher ist das resultierende System sowohl zuverlässig als auch robust und stellt somit eine ausreichende Energieversorgung sicher. Aspekte der Nachhaltigkeit werden durch die Anwendung multikriterieller Optimierung berücksichtigt. Zur Entscheidungsunterstützung wählt die Methode automatisch ein einziges Design aus. Das ausgewählte Design ist äußerst flexibel, sodass sich der Betrieb bezüglich der betrachteten Kriterien anpassen lässt.

Die vorgestellte Methode wird auf eine reale Fallstudie angewandt, in der das Design eines dezentralen Energieversorgungssystems bezüglich der jährlichen Gesamtkosten sowie der Treibhausgas-Emissionen optimiert wird. Die Ergebnisse zeigen, dass sowohl Zuverlässigkeit als auch Robustheit schon mit geringen Mehrkosten erzielt werden können. Die Methode wählt ein Design mit höherer Flexibilität im Betrieb aus als vergleichbare Designs bieten, die durch die alleinige Anwendung von multikriterieller Optimierung gefunden werden. Die Ergebnisse zeigen, dass die vorgestellte Methode ein praxistauglicher Ansatz ist, um ein höchst flexibles nachhaltiges Systemdesign auszuwählen, das gleichzeitig zuverlässig und robust ist.

Abstract

The synthesis of energy systems is a complex task, since a plethora of conditions needs to be regarded during decision making. Thus, mathematical optimization is an excellent tool to accomplish this task and to identify an optimal system design. However, energy system synthesis is intrinsically uncertain, since the availability of components is inherently uncertain as well as the input parameters, such as energy demands. As a result, neglecting uncertainties might lead to a lack of energy supply. An insufficient energy supply possibly causes both high unexpected costs and environmental damage. Thus, uncertainties need to be regarded during optimization of energy systems. At the same time, sustainability is a further major aspect in the synthesis of energy systems. To regard sustainability performance, multiple decision criteria, such as economic, environmental, and social criteria, need to be taken into account. For this purpose, employing multi-objective optimization is perfectly suitable. However, multi-objective optimization leaves the decision maker in general with more than one solution to choose from which is often very challenging.

Therefore, in this thesis, a framework to select the **be**st **re**liable and ro**bu**st **s**us**t**ainable design of an energy system is proposed – the *be-rebust framework*. The framework takes into account both, uncertainty of energy supply as well as uncertainty of input parameters for optimization. Thus, the designed system is reliable and robust guaranteeing security of energy supply. Sustainability is regarded by employing multi-objective optimization. For decision support, the framework automatically selects one single design. The selected design allows for highly flexible operation regarding the considered objective criteria.

The proposed be-rebust framework is applied to a real-world case study. In the case study, the design of a distributed energy supply system is optimized. For this purpose, total annualized costs and the global warming impact are minimized. The results show that reliability as well as robustness can be achieved with only low additional costs. Employing the framework enables to select a sustainable design with higher operational flexibility than provided by designs identified by sole application of multi-objective optimization. The results verify the excellent performance of the proposed framework to select the best sustainable energy system design which is reliable and robust.

Contents

List of Figures

List of Tables

Nomenclature

Latin symbols

INV	total investment costs
D	distance measure
$\dot{E}$	energy demand
GWI	(specific) global warming impact
I	investment costs of components
K	number of failing components
$\mathcal{MP}$	multi-objective optimization problem
m	gradient
n	number of components
N	number of points on Pareto front
$\mathcal{OP}$	optimization problem
$OPEX$	annualized operational costs
$\mathcal{P}$	Pareto front
p	price factor
pg	relative variation of uncertain gas price
pe	relative variation of uncertain electricity price
PVF	present value factor
r	interest rate
TAC	total annualized costs
$\dot{U}$	input energy flow
$\dot{V}$	output energy flow
y, z	auxiliary variables for MILP reformulation

Greek symbols

α	auxiliary variable to limit objective function
δ	variation of uncertain values
$\Delta\tau$	duration
ε	auxiliary variable for minimizing the distance of Pareto fronts
γ	binary variable for linearization
ϑ	time horizon
κ	specific investment costs for linearization
λ	weighting factor
η	efficiency/ coefficient of performance
ω	uncertainty-scaling factor
ρ	point on ideal Pareto front
ξ	scenario
ζ	scenario to describe failure

Super- and subscripts

$\widehat{\ }$	nominal value
$\tilde{\ }$	uncertain value
$\overline{\ }$	upper bound of parameters and normalized objective function
$\underline{\ }$	lower bound
$\lvert\ \rvert$	cardinality
*	optimal
A	Site A
B	Site B
buy	purchased
c	cooling
$\dot{E}h/c/el$	related uncertain demands
el	electric
f	first stage
gas	natural gas
h	heating
$ideal$	ideal

(continues on next page)

lb	lower bound
m	maintenance
max	maximal
N	installed (related to components)
nom	nominal
$\mathcal{OW}$	objective wise
$\mathcal{RC}$	robust counterpart
$rely$	reliable
rr	reliable and robust
rob	robust
pg	related to uncertain gas price
pe	related to uncertain electricity price
s	second stage
$sell$	sold
T	transpose
tot	total
u	dimension of the uncertainty set
wc	worst case

Elements and sets

$[Z]$	corresponds to $\{1, \ldots, Z\}$ with any integer $Z \in \mathbb{N}$
$h \in [H]$	index of line segment
$i \in [L]$	index of objective functions
$j, l \in [N]$	index of points on Pareto front
$k \in \mathcal{AC}$	absorption chiller
$k \in \mathcal{B}$	boiler
$k \in \mathcal{CC}$	compression chiller
$k \in \mathcal{CHP}$	combined heat and power engine
$k \in \mathcal{K}$	component of the energy system with $\mathcal{K} = \mathcal{B} \cup \mathcal{CHP} \cup \mathcal{AC} \cup \mathcal{CC}$
$k' \in \mathcal{K}' \subseteq \mathcal{K}$	failing component
$t \in \mathcal{T}$	time step
$\xi/\zeta \in \mathcal{U}$	scenario of the uncertainty set
$x \in \mathcal{X}$	feasible solution

Abbreviations

CHP	combined heat and power
CPU	central processing unit
DESS	distributed energy supply system(s)
e. g.	exempli gratia (English: for example)
Eq.	equation
Fig.	figure
flex-hand	flexible here-and-now decision
i. e.	id est (English: that is)
MILP	mixed integer linear program
RAM	random-access memory
be-rebust	best reliable and robust sustainable
RoMO	Robust Multi-objective Optimization
Tab.	table
TRusT	Two-stage Robustness Trade-off

Chapter 1

Introduction

Availability of clean, affordable, and reliable energy has been indispensable for economic growth since the industrial revolution (Chu and Majumdar, 2012). Today and in the future, sustainability is an additional major aspect in the synthesis of energy systems (Demirhan et al., 2019). The reduction of global greenhouse-gas emissions is the key to achieve the goal of the Paris Agreement (2015) of limiting the global temperature increase to a maximum of 2 °C compared to the pre-industrial temperature level. The executive director of the International Energy Agency has summarized the challenge:

> *Strong policies and innovation can make the difference for energy security, climate change, air quality, and universal access to modern energy services in parallel – in short, building a secure, affordable, sustainable energy system that is available to all.*
>
> Fatih Birol (2018)

Policy decisions are one important element for implementing sustainable energy systems. Innovations provide the foundation for policy decisions (Bogenschneider and Corbett, 2011). Hence, innovations are required for significant reductions of global greenhouse-gas emissions of energy systems or even for a climate-neutral energy sector.

> *Achieving climate neutrality in the energy sector – while ensuring at the same time a more efficient energy use, a secure supply of energy, affordable prices and low environmental impact – is a complex endeavor.*
>
> European Commission (2018)

The quotes emphasize that designing a sustainable energy system which guarantees security of energy supply is a *challenging* task but also a *necessary* task.

Energy systems are more frequently designed using mathematical optimization models (Andiappan, 2017). Often, deterministic optimization models are employed (Mancarella, 2014). However, deterministic optimization models rely on the assumption of perfect foresight. Perfect foresight implies that input parameters (e. g., energy demands or prices for gas and electricity) are known with certainty when the optimization is performed. Furthermore, availability of all energy supply components is often assumed in optimization models. If the actual conditions differ from the assumptions considered during optimization, the obtained solutions will usually become suboptimal or even infeasible, as shown for general optimization problems by Ben-Tal and Nemirovski (2000) and specifically for distributed energy supply systems by Sun et al. (2017). For energy systems, suboptimality and infeasibility often imply insufficient energy supply. Insufficient energy supply might occur, if an energy supply component fails or if the actual energy demand is higher than the maximal assumed demand during synthesis of the energy system. As a result, further components or production units might fail which possibly leads to extremely high costs or environmental damage. Thus, the uncertainty of parameters and the uncertainty of component availability needs to be taken into account already during optimization (DeCarolis et al., 2017; Moret, 2017).

Designing energy systems which are not only secure but which also regard aspects of *sustainability* further increases the problem complexity. Optimization of sustainable energy systems requires multiple criteria to quantify sustainability performance. In this thesis, the expression *sustainable energy system* refers to energy system designs with highest possible sustainability performance regarding specific criteria of sustainability. The employed criteria to measure sustainability performance comprise three important dimensions: economy, ecology, and society (Kloepffer, 2008). Thus, the synthesis of a sustainable energy system is intrinsically a multi-objective optimization problem. Multi-objective optimization usually leads to various solutions which cannot be improved regarding one criterion without worsening at least one other criterion (Ehrgott, 2005). In consequence, the designer of the energy system has to answer the question: *Which design should I choose?*

This thesis aims to get one step closer to "*[...] climate neutrality in the energy sector – while ensuring at the same time a more efficient energy use, a secure supply of energy, affordable prices and low environmental impact [...]*" (European Commission, 2018) by proposing a framework to select the **be**st **re**liable and ro**bust** **s**ustainable design of an energy system – the *be-rebust framework*.

The remaining thesis is structured as follows: Chapter 2 provides an overview about the current state of the art in energy system optimization. The resulting open research question and the contribution of this thesis is stated in Chapter 3. Chapter 4 provides the employed optimization model with the corresponding uncertainties, as well as the considered case study of this thesis. In Chapters 5-7, the developed mathematical approaches are introduces. The approaches are assembled to one framework in Chapter 8, to answer the open research question. A summary, conclusions, and future perspectives are provided in Chapter 9.

Chapter 2

State of the art in energy system optimization

This chapter reviews the state of the art in optimal synthesis of energy systems. First, characteristics of *distributed energy supply systems (DESS)* are introduced (Section 2.1). The focus of the remaining review is on optimization approaches for energy systems, including two-stage optimization (Section 2.2), multi-objective optimization (Section 2.3), and optimization under uncertainty (Section 2.4).

2.1 Distributed energy supply systems (DESS)

The industry sector contributed around 37 % to the total global final energy demand in 2017 according to the International Energy Agency (Fernandez-Pales et al., 2019). In general, industrial sites require several forms of energy, in particular, high heating and cooling demands besides electricity demands. To cover these demands, a combined energy generation is often beneficial, since 10 % up to 30 % of primary energy can be saved (Pepermans et al., 2005). DESS are commonly tri-generation systems and thus allow for combined generation of multiple energy forms. Besides industry, distributed generation is useful for energy supply in research campuses (Sass et al., 2020), urban districts (Weber, 2008; Jennings et al., 2014; Jing et al., 2019b), and buildings (Schütz et al., 2017). Hence, this thesis focuses on DESS. However, the approaches proposed in this thesis can be extended to general energy systems or even more general problems as discussed in the corresponding chapters.

DESS often employ combined cooling, heating, and electricity systems in order to maximize their efficiency and to decrease greenhouse-gas emissions (Chicco and Mancarella, 2009; Weber, 2008): To cover heating demands, DESS typically include boilers and/or combined heat and power engines both running on fuel. Beyond supplying heat, combined heat and power engines, such as gas turbines, also provide electricity. The provided heat is usually used on site to cover heating demands. Furthermore, the heating system is often coupled to the cooling system: Absorption chillers transform heating into cooling energy (Minciuc et al., 2003). Such trigeneration systems are often considered in literature when optimizing energy systems (Ünal et al., 2015). To alternatively cover cooling demands, the trigeneration system can be extended by installing electrical chillers, such as compression or turbo chillers (Minciuc et al., 2003; Ziher and Poredos, 2006) which can be driven by electricity from the combined heat and power engines or from the electricity grid. Besides these classical components, the possibly installed design of DESS can be broadened by a wide variety of component types such as storage systems, heat pumps, organic Rankine cycles, and renewable energy conversion components as wind turbines, photovoltaic units, and solar heat panels (Baumgärtner et al., 2019b).

2.2 Two-stage optimization of energy systems

The synthesis of energy systems can also be interpreted as a *two-stage optimization problem* (Lin et al., 2016). Two-stage optimization problems are characterized by two kinds of variables (Ben-Tal et al., 2004): *First-stage variables* have to be fixed in the beginning (so-called *here-and-now variables*); *second-stage variables* (*wait-and-see variables*) can be revised later when knowledge on the respective scenario is available. A *scenario* represents one combination of parameters that might occur. In the synthesis of energy systems, Frangopoulos et al. (2002; 2018) have introduced three levels of decision making to optimize energy systems corresponding to structure, sizing, and operation (Voll et al., 2013). However, these three levels can be condensed into two stages (Fig. 2.1): The first-stage variables x^f commonly correspond to design variables determining the structure of the energy system and the sizing of the components. The second-stage variables x^s correspond to operational variables defining the operation of the installed components.

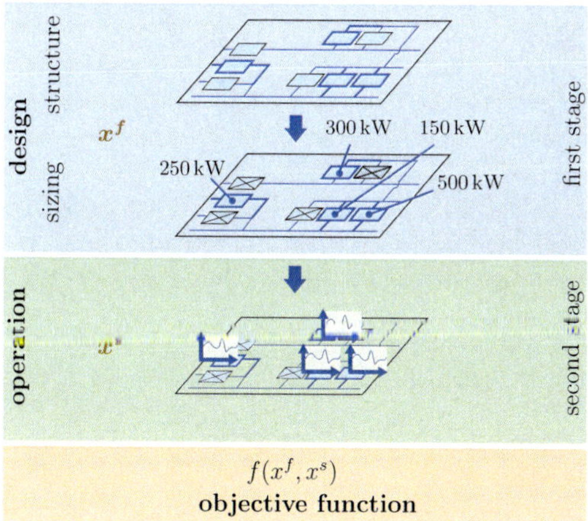

Figure 2.1: Two stages of energy system optimization (Lin et al. (2016); design variables x^f and operational variables x^s) including three levels (structure, sizing, and operation; Frangopoulos et al. (2002))

A general two-stage problem formulation is given by

$$\begin{aligned} \min\ & f(x^f, x^s) \\ \text{s. t.}\ & x^f \in \mathcal{X}^f \\ & x^s \in \mathcal{X}^s(x^f)\,. \end{aligned}$$

In the optimization of energy systems, the objective function $f(x^f, x^s)$ typically represents economic criteria, such as total annualized costs combining investment and operational costs. The set of feasible solutions $\mathcal{X}^s(x^f)$ for the second-stage variables x^s depends on the feasible first-stage solution $x^f \in \mathcal{X}^f$. The sets of feasible solutions $\mathcal{X}^f$ and $\mathcal{X}^s$ are determined by constraints which need to be regarded during optimization. In general, for energy system optimization, these constraints typically include:

- energy balances,
- design and operational limits, such as limited component sizing or minimal part-load restrictions, and

- specifications of the energy conversion components, such as part-load performance.

Further problem-specific constraints can be regarded, e. g., for considering a peak-power price (Baumgärtner et al., 2020).

The synthesis of DESS involves non-linearities, e. g., due to part-load performance and, if considered, due to investment costs (Bruno et al., 1998; Goderbauer et al., 2016). The resulting mixed-integer non-linear problem are usually challenging for standard solvers. Thus, tailor-made algorithms are often developed, for example, by Goderbauer et al. (2016) and by Beykal et al. (2018). However, frequently, mixed-integer linear problems (MILP) are used (Mancarella, 2014; Andiappan, 2017) with reformulated non-linearities, e. g., using piece-wise linearization (Voll et al., 2013).

In energy system optimization, the two-stage nature is often exploited to decrease computational effort. For example, computational effort is decreased by using heuristic approaches on the first stage to find promising design candidates, e. g., employing evolutionary algorithms (Elsido et al., 2017) or meta-heuristic algorithms (Wakui et al., 2019). Based on these design candidates, the operational optimization problem is solved with an exact approach. Another way to decrease computational effort is to decrease the model size on the first stage by aggregated time series (Bahl et al., 2018a,b; Baumgärtner et al., 2019a; Yokoyama et al., 2019) while using the full time series only on the second stage. Further advantage of the two-stage nature is taken in multi-objective optimization (Section 2.3) as well as in optimization under uncertainty (Section 2.4.2).

2.3 Multi-objective optimization and design selection of sustainable energy systems

In the optimal synthesis of energy systems, multi-objective optimization is often employed to consider aspects of sustainability. An optimization problem regarding L objective functions $f_i(x)$ with $i \in \{1, \ldots, L\}$ is given by:

$$\begin{aligned} &\min \left(f_1(x), \ldots, f_L(x)\right)^T \\ &\text{s.t. } x \in \mathcal{X} . \end{aligned}$$

The superscript T is used to denote the transpose and $\mathcal{X}$ represents the set of feasible solutions. A solution is called *efficient* (also *Pareto efficient*), if the objective value regarding one objective function cannot be improved without simultaneously worsening the objective value of another objective function. The set of objective values corresponding to the set of efficient solutions is called *Pareto front*. If a solution is not efficient, the solution is *dominated* by another solution which has a lower objective value in at least one criterion while all other objective values are not worse. For a detailed introduction to multi-objective optimization, see Ehrgott (2005).

In this thesis, we[1] refer to *sustainable energy systems* as energy system designs leading to objective function values on the Pareto front of a multi-objective optimization regarding aspects of sustainability. Hence, sustainable designs have highest possible sustainability performance (Chapter 1). For the optimization of sustainable energy systems, we identify the following typical objective functions in current literature based on the literature review of Hennen et al. (2017): Gebreslassie et al. (2012) and Salcedo et al. (2012) consider the trade-off between total annualized costs and global warming impact. Similar criteria are used by Flores et al. (2015) and Giarola et al. (2011) considering the net present value and greenhouse-gas emissions. Weber (2008), Fazlollahi et al. (2012), and Buoro et al. (2013) investigate total annualized costs and CO_2 emissions. Fazlollahi et al. (2014a) additionally add the efficiency of the energy system as third objective function. First articles on social criteria have been published (Ramos et al., 2014; Mota et al., 2015; Popovic et al., 2018). However, most concepts of sustainable system design focus on economic and environmental sustainability (for reviews see Pohekar and Ramachandran (2004) as well as Grossmann and Guillén-Gosálbez (2010)). Hence, in this thesis, we focus on economic and environmental criteria (Section 4.1.1).

When employing multi-objective optimization for the design of energy systems, the decision maker receives not only one optimal design but a whole Pareto front with many different designs. Hence, the decision maker is often confronted with the question: *How to select one single design?*

In literature, several approaches exist to reduce the number of relevant solutions, so that the decision maker has to choose from fewer options. A method to focus on the relevant solutions *a priori* by excluding solutions before the optimization is proposed by Branke et al. (2004) and Rachmawati and Srinivasan (2009). They

[1]This thesis is written in the pluralis modestiae to avoid the excessive use of passive voice.

introduce a preference-based evolutionary approach focusing on calculating "knee" regions of the Pareto front. Vallerio et al. (2015) propose an interactive decision-support system for multi-objective optimization of nonlinear dynamic processes with uncertainty. Another interactive tool is introduced by Bortz et al. (2014) allowing the decision maker to navigate along the Pareto front to focus on the relevant solutions. Hennen et al. (2017) focus on Pareto-efficient solutions which are near-optimal with respect to an aggregated criterion that represents the overall set of objective functions.

An *a posteriori* approach to reduce the set of relevant Pareto-efficient solutions is to cluster solutions, e. g., based on subtractive clustering (Zio and Bazzo, 2011), by k-means classification (Taboada et al., 2007), or by a self-organizing map (Li et al., 2009). Das (1999) focuses on relevant Pareto-efficient solutions by evaluating subsets of the objective functions. The approach has been further developed by Antipova et al. (2015) applying Pareto filters. But still, in all approaches which reduce the number of relevant solutions, the decision maker has to select the finally implemented design.

For further reduction of the Pareto-efficient solutions, Taboada et al. (2007) and Abubaker et al. (2014) propose ranking methods which are based on a prioritization of objective functions by the decision maker. For ranking compromising solutions without explicit prioritization, the methods LINMAP (linear programming technique for multidimensional analysis of preference; Srinivasan and Shocker (1973)), VIKOR (Duckstein and Opricovic, 1980; Opricovic and Tzeng, 2004), and TOPSIS (technique for order preference by similarity to an ideal solution; Hwang and Yoon (1981); Chen and Hwang (1992)) measure the distance from the ideal point and, in TOPSIS, additionally from the nadir point. The *ideal point* is assembled by the respective single-objective optimal values of the objective functions, the *nadir point* by the worst possible values. The idea of the 3 approaches is based on compromise programming, where the solution with objective values "as close as possible" to the ideal point is chosen, or "as far away as possible" from the nadir point (Zelany, 1974). With the aim to determine key players in social networks, de la Fuente et al. (2018) employ eleven methods for automatic solution selection within the set of Pareto-efficient solutions. These include, e. g., shortest distance to the ideal point (Padhye and Deb, 2011), shortest distance to all points, highest hypercube (Beume et al., 2009), and consensus (Pérez et al., 2017). However, non of the approaches takes advantage of two-stage characteristics. Single-stage methods for a posteriori decision making have been recently reviewed by Jing et al. (2019a). For general multi-objective problems, the ranking methods reviewed in this section are suitable to select one solution. However, for two-stage problems, as the synthesis of energy systems (Section 2.2), the reviewed approaches could miss well-performing solutions as the approaches depend on pre-calculated efficient solutions. Common solution algorithms for multi-objective optimization problems only calculate a discrete set of the Pareto front. Thereby, well-performing designs might not be part of the pre-calculated Pareto front and thus not taken into account for final selection.

The approaches for solution reduction discussed so far do not take any advantage of the two-stage characteristics of energy systems. Thus, these approaches miss the opportunity to choose a first-stage solution which provides high flexibility on the second stage. Exploiting flexibility might be particular important, since future conditions are not known today. Hence, the capability of energy systems to adapt operation to changing circumstances should be targeted (Shang and Kokossis, 2005). Political aims might change; and thus, the importance of economic, environmental, and social aims might change. As a result, sustainable energy systems should provide flexible operation to enable adaptation to a changing focus within the regarded criteria.

Two-stage characteristics are regarded in design selection by Mattson and Messac (2003). In their approach, the design options need to be discrete. For each design option, the corresponding Pareto front is generated. Afterwards, a Pareto filter is applied simultaneously to all generated fronts deleting all dominated solutions. Finally, design options lying inside a pre-defined region of interest are selected. However, with increasing number of considered components during optimization (called *superstructure*), the design options along the Pareto front might change more frequently within the region of interest. Hence, a higher number of suitable design options needs to be taken into account during final selection. A similar approach is proposed by Carvalho et al. (2012): Based on discrete design options, the corresponding Pareto front is generated. Design options with the ability to undergo large changes in operation enable higher resilience and are thus favored. Guo et al. (2013) introduce a two-stage optimal planning and design method for combined cooling, heating, and power microgrid systems. On the first stage, the system design is optimized using a generic multi-objective optimization approach. For several obtained designs, the operational costs are minimized on the second stage. Wang et al. (2019) additionally regard a feedback from the second stage to the first stage to ensure the accuracy of the planning. However, a systematic or even automatic selection of favored design options is not proposed in either approaches, e. g., by applying distance measures to assess the generated Pareto fronts.

All discussed approaches do *either* take advantage of the two-stage characteristic of energy systems *or* propose a ranking of Pareto-efficient solutions. How to select one single design while exploiting the two-stage nature has not been proposed so far to the author's best knowledge.

2.4 Uncertainties in energy system optimization

In the optimal design of energy systems, the synthesis depends on many input parameters which are inherently uncertain, such as future energy demands or prices. However, not only input parameters are uncertain but also the availability of components is uncertain (Aguilar et al., 2008). The following Sections 2.4.1 and 2.4.2 provide a detailed overview of how to handle uncertainties in energy system synthesis.

2.4.1 Uncertainty of energy supply

If a component of an energy system fails and is not available, the energy demand has to be covered by the remaining components of the system. If the remaining energy system is not able to supply the demanded amount of energy, the energy system is not reliable. In literature, *reliability* is defined as the probability that a system is able to provide a required function; *availability* describes the probability that an item delivers its required function during a certain time period (Aguilar et al., 2008). Since a lack of energy supply can only be avoided with certainty if the reliability of the system is 100 percent, a system with 100 percent reliability is regarded as a reliable system in this thesis.

In practice, reliability is often aspired by rules of thumb where additional units are added to the optimal energy system (Aguilar et al., 2008). Rules of thumb and other heuristics usually result in a suboptimal design, since the additional components are not considered in the optimization (Andiappan et al., 2015). Therefore, developing mathematical concepts for reliability is an important research area.

To identify reliable designs, several stochastic optimization approaches have been proposed. Sun and Liu (2015) propose a multi-period stochastic programming approach to design steam power systems. They take into account both uncertain availability of components and uncertain demands to obtain reliable and robust energy systems. Frangopoulos and Dimopoulos (2004) investigate the effect of failing components on the design of the energy system and the operation of the components. They employ the state-space method (for a brief introduction see Frangopoulos and Dimopoulos (2004)) to simulate partial failure of the system. Also based on the state-space method, Miryousefi Aval et al. (2015) use a two-state Markov chain to design reliable building cooling, heating, and power systems and analyze the system's impact on the existing electrical power system. Abdollahzadeh and Atashgar (2017) propose a bi-objective two-stage stochastic programming model to obtain an optimal design regarding uncertainty of energy supply, a maintenance strategy, and inspection intervals for supplier systems (e. g., for a wind farm). For this purpose, they minimize the system life-cycle costs and maximize the system availability. Costs and reliability of the system are also optimized by Jahromi and Feizabadi (2017) to solve the redundancy allocation problem (Barlow et al., 1965) with the aim of installing parallel redundant components. They use a gamma distribution to describe component reliability. In order to analyze the trade-off between system economics and system failure, Ji et al. (2014) introduce risk preferences of decision makers in stochastic approaches. A similar trade-off between costs and availability is considered by Ye et al. (2017). Their proposed MINLP selects the optimal design of a reliable serial system comprising parallel components. In the problem formulation, the availability of the considered chemical process units is also described via probabilities. Recently, Ye et al. (2019) have extended this framework and introduced a systematic approach to regard the failure and repair process using a Markov chain. Andiappan et al. (2015) propose an approach based on k-out-of-n system modeling (Birnbaum, 1968) to find reliable biomass-based tri-generation systems. A k-out-of-n system is still reliable if k out of n components are still available. In their approach, Andiappan et al. (2015) determine redundancy allocation of process units with a specified minimum reliability level using chance-constraint programming. Andiappan and Ng (2016) extend this k-out-of-n approach employing a generic formulation to incorporate operation strategies for the design of reliable tri-generation systems. However, employing stochastic approaches might still lead to a lack of energy supply, since probabilities are regarded in the optimization. Thus, stochastic approaches do not lead to a 100 percent reliable system.

To avoid stochastic approaches, scenario-based analysis has been considered: Andiappan et al. (2017) propose an MILP to analyze the effects of unavailability of an energy conversion unit on the system flexibility without employing stochastic optimization. Instead, they use input-output modeling (Leontief, 1936) based on disruption scenarios to identify insufficient flexibilities and deduce a step-by-step guide to retrofit an existing energy system. Another rigorous algorithm, also based on scenarios, has been proposed by Caserta and Voß (2015). They incorporate redundancies by transforming the reliability redundancy allocation problem into a multiple-choice knapsack problem and employ a multi-period approach to take potential scenarios into account. Aguilar et al. (2008) consider maintenance and failure scenarios to design reliable energy supply systems regarding the largest one, two, or more components to fail. Thus, for the considered scenarios, the resulting energy supply system is able to provide the necessary amount of energy even if a component fails during maintenance of other components. Considering scenarios of failure implies that failure of any component not included in the scenarios might still lead to a lack of energy supply. Thus, trustworthiness of reliability calculations always depends on the correct selection of scenarios.

In power systems engineering, reliable electricity supply is crucial and an active research area. Thus, many approaches have been proposed to increase reliability employing optimization: Ruiz and Conejo (2015) and Mínguez and García-Bertrand (2016) employ adaptive robust optimization to solve the transmission expansion planning problem. A scenario-based approach is proposed by Alguacil et al. (2009) evaluating the trade-off between investment cost reduction and vulnerability of the transmission network against attacks using a weighted objective function. Ruiz et al. (2009) enforce reserve requirements in a stochastic formulation to compensate the limited representation of uncertainty by the selection of scenarios. In their paper, uncertainty refers to both load uncertainty and supply uncertainty. Choi et al. (2005) propose a methodology based on probabilistic reliability criteria to minimize cost of transmission system expansion. In a later paper, Choi et al. (2006) additionally regard costs of outages. Another well-known approach in power systems engineering is $(n-K)$-reliability. $(n-K)$-reliability ensures reliability of a power system with n components during the failure of maximal K components. For K being 1, stochastic approaches have been developed to find $(n-1)$-reliable solutions for network expansion and transmission switching in reasonable time, e. g., by Hedman et al. (2010) and Wiest et al. (2018). For the German power grid, $(n-1)$-reliability represents an adequate reliability of supply (Berndt et al., 2007). Employing robust optimization, $(n-K)$-reliability is employed in contingency-constrained transmission expansion planning by Moreira et al. (2015) and additionally regarding load uncertainty by Hong et al. (2017). In contingency-constraint unit commitment, Wang et al. (2013) and Street et al. (2011) introduce $(n-K)$-reliability using a robust formulation. Street et al. (2011) propose a rigorous approach which does not depend on the cardinality of the uncertainty set. However, this approach is based on the assumption of single-bus unit commitment problem in which only upward reserves need to be considered. This shortcoming has been overcome by Street et al. (2014) who also take into account failure in the transmission network. To solve the extended model, a Benders' decomposition approach (Benders, 1962) is employed.

The concept of reliability is also crucial for DESS. In DESS, several energy conversion components are usually operated to supply the desired energy demand. Failure of any component could always occur and the remaining components would need to compensate the loss to ensure reliable operation. However, the approaches from power system engineering for $(n-K)$-reliable transmission expansion planning cannot be directly adapted to DESS. In transmission networks, only electricity is considered. In contrast, in DESS, multiple types of final energy need to be taken into a count. A coupling between circuits of different energy types increases the complexity of the synthesis problem, since a failure of a component in one circuit might affect another circuit. Hence, tailor-made approaches for DESS are required.

2.4.2 Uncertainty of input parameters

If the actual conditions differ from the deterministic parameters considered in the optimization, the obtained solutions will usually become suboptimal or even infeasible, as shown for general optimization problems (Ben-Tal and Nemirovski, 2000) as well as for typical engineering problems, such as unit commitment (Shi and You, 2016), production planning (Li et al., 2011), and DESS (Sun et al., 2017).

For energy systems, infeasibility implies insufficient energy supply. For instance, if uncertain energy demands are not taken into account during the planning of DESS, a lack of energy supply may arise for the designed system. Not only demands are uncertain when designing sustainable energy systems but further uncertainties influence the decision, e.g., the future electricity mix, policy changes, or energy prices. Thus, uncertainties have a significant impact and have to be taken into account in the optimization (Akbari et al., 2014).

In practice, a pragmatic approach is often employed to aim for a robust energy supply: Peak loads are taken into account in the optimization when considering average monthly values (Voll et al., 2013), typical days (Domínguez-Muñoz et al., 2011) or other typical periods (Fazlollahi et al., 2014b; Bahl et al., 2018b; Kotzur et al., 2018). While sufficient energy supply can be achieved still assuming perfect foresight but with extraordinary high peak demands, the resulting system might be suboptimal in operation as the peak demands may not lead to the worst possible costs. Thus, the uncertain nature of demands has to be considered in the optimization to identify a cost-efficient robust energy system. The definition of a *robust energy system* is not used consistently in literature (Månsson et al., 2014). We expect a robust energy system to cover uncertain energy demands but without regarding failure of equipment.

If probability distributions of the uncertain parameters are available, stochastic optimization (Birge and Louveaux, 2011) can be applied. Guzman et al. (2016, 2017) propose a priori and a posteriori bounds for uncertain parameters with unknown and attributed known distributions. Ning and You (2017) extract probability distributions from uncertain parameters to avoid over-conservatism. Amusat et al. (2017) quantify stochastically the effect of uncertain weather data on the design of renewable energy systems. In literature, various stochastic approaches for optimization of energy systems exists (Grossmann et al., 2016); however, probability distributions are unknown in general and have to be estimated. The uncertainty in the estimated probability distribution then impacts on the actual robustness of the design (Delage and Ye, 2010).

Robust optimization overcomes the need for uncertain probability distributions. A robust solution considers every possible scenario without relying on its probability of occurrence. Robust optimization is a very active research field with a wide range of approaches (a detailed introduction is given by Goerigk and Schöbel (2016)). In the following, an overview is provided on the major current research relevant to energy systmes including strictly robust optimization, Γ-robustness, minmax regret, and adjustable robustness.

Strictly robust optimization (Soyster, 1973) (also called *minmax robust optimization*) ensures the feasibility of a solution for every considered scenario while minimizing costs for the most expensive scenarios. The concept of strictly robust optimization was introduced for linear programs by Soyster (1973) and restated by Ben-Tal and Nemirovski (1999). The resulting strictly robust optimal solutions are in general very conservative and, thus, expensive compared to the nominal costs. The *nominal costs* represent costs obtained when uncertainties are neglected and perfect foresight is assumed.

In order to reduce the degree of conservatism of robust solutions, the so-called Γ*-robustness* was introduced by Bertsimas and Sim (2004). In Γ-robustness, not all uncertain parameters are assumed to vary at the same time. The level of uncertainty Γ limits the number of parameters varying simultaneously and represents the degree of conservatism. For linear problems, the resulting Γ-robustness program can be transformed into a linear program. Robust optimization approaches have been applied successfully to the operation of DESS: Dong et al. (2013) use the Γ-robustness to derive a fuzzy programming model for planning robust energy management systems with environmental constraints. Akbari et al. (2014) employ Γ-robustness to optimize operation. Additionally, they determine the Γ-robust structure and sizing of the installed heating and cooling components, considering renewable energy and storage. Renewable energy technologies are also considered in the design of energy systems by Moret et al. (2014, 2019). Moret et al. (2014) classify the uncertain parameters to define adequate ranges of their variation. Recently, Moret et al. (2019) have proposed a decision support method employing Γ-robustness for energy planning. The decision support is based on the idea "first feasibility, then optimality". In general, Γ-robustness can only ensure sufficient energy supply if the conservativeness level Γ is set to its upper bound. However, using the upper bound for the level Γ leads to minmax robust optimization and thus to expensive solutions which are often not relevant for practical applications. Reducing the degree of conservatism in any way, however, compromises security of energy supply. This compromise can be measured by a robustness index analyzing the trade-off between investments and achievable targets (Sy et al., 2016).

The *minmax regret approach* (e. g., Yaman and Pinar (2001)) aims to identify a solution with low risk, characterized by the so-called *regret*. For each scenario, the regret is defined as the deviation between the occurring costs and the minimal possible costs for the scenario. The maximal possible regret is then minimized. The minmax regret approach is successfully applied by Dong et al. (2011) to determine power generation and capacity expansion for uncertain demands. Yokoyama et al. (2014, 2018) take advantage of the two-stage nature of energy systems and introduce a three-level approach regarding hierarchical relationships among design variables, uncertain energy demands, and operation variables. To decrease the conservatism in minmax regret, Yokoyama and co-workers allow a violation of the energy balance constraints and introduce a penalty term for the unmet energy demands into the objective function.

The concept of two stages has also been introduced into robust optimization as *adjustable robustness* (also called *adaptive robustness*) (Ben-Tal et al., 2004). The original concept was further extended by Thiele et al. (2010). For a current survey see Yanıkoğlu et al. (2019). Bertsimas et al. (2013) apply adaptive robustness to a unit-commitment problem regarding the uncertainty of energy supply as well as uncertain loads. Unit commitment is considered as first stage and dispatch as second stage. Adaptive robustness can be combined with a conservatism scaling factor to achieve less conservative solutions Shi and You (2016). Applying adaptive robustness leads to complex problems which are hard to solve. Thus, special solving strategies are necessary, such as Bender's decomposition (Benders, 1962) and outer approximation (Duran and Grossmann, 1986). A detailed discussion on two-stage programming methods for process systems under uncertainty is provided by Grossmann et al. (2016).

The literature review in this section shows that most robust design approaches aim to be less conservative than minmax robust optimization, in order to obtain less expensive solutions. In return, however, they cannot guarantee sufficient energy supply.

Recently, the idea of robustness has been extended to multi-objective optimization: *Minmax robust multi-objective optimization* is based on the classical concept of minmax robust optimization. In minmax robust multi-objective optimization, multiple uncertain objectives are minimized while ensuring feasibility of the solution for every scenario (Kuroiwa and Lee, 2012; Ehrgott et al., 2014; Fliege and Werner, 2014; Goberna et al., 2015; Bokrantz and Fredriksson, 2017; Botte and Schöbel, 2019; Dranichak and Wiecek, 2019). The combination of robustness with multi-objective optimization is a young and very active research area. First successful applications of minmax multi-objective robustness have been realized in the fields of internet routing (Doolittle et al., 2018), portfolio optimization (Fliege and Werner, 2014), and aircraft routing (Kuhn et al., 2016). Other approaches regarding uncertainties have also been extended to the multi-objective case, e. g., by Vallerio et al. (2016) combining a stochastic approach with multi-objective dynamic optimization for chemical vapor deposition, and by Deb and Gupta (2004) who exclude uncertain optima by smoothing the objective functions. Asprion et al. (2017) and Bortz et al. (2017) extend the approach to focus on relevant solutions (proposed by Bortz et al. (2014); Section 2.3) by accounting for uncertainties in process optimization. Ide and Schöbel (2016) provide a survey and an analysis of approaches for one-stage robust multi-objective problems. For DESS optimization, Karmellos et al. (2019) provide a comparison of methods regarding uncertainty of input parameters.

Still, for multi-objective problems with uncertainty of input parameters, the same question arises as for deterministic problems: *How to select one single design?*

An evolutionary approach is proposed by Quintana et al. (2017) which generates only solutions which are more tolerant to deviations. However, not only one but a set of solutions is generated and decision makers' preferences are necessary as well. To rank solutions, Lemos et al. (2018) discriminate Pareto-efficient solutions by their sensitivity against process uncertainties. For this purpose, they minimize the distance between disturbed results and the original Pareto front. However, uncertainty is not regarded in the generation of the Pareto-efficient solutions. Karimi and Jadid (2019) regard uncertainties by minimizing unfilled loads and maximal reserves provided by the storage units. As third objective, costs are selected to schedule microgrids. A ranking of solutions is provided but needs decision makers' preference as input. Pulsipher et al. (2019) use a flexibility index describing the ability of the system to adapt to the largest uncertainty. The flexibility index is used to rank design options. Additionally, their approach identifies flexibility-limiting constraints. However, the flexibility index is only used for assessing given design options and is not incorporated in the optimization of the design. For an analysis of the relation between flexibility analysis and adaptive robustness see Zhang et al. (2016). Tock and Maréchal (2015) rank Pareto-efficient solutions employing uncertainty analysis assessing the probability of the solution to perform the best, given a chosen performance indicator. Sun et al. (2018) identify a single solution by converting multiple objective functions into one unified cost function. However, the aggregation to only one function already assumes a weighting of objective criteria by the decision maker.

When taking uncertainties into account during minimization of multiple criteria, two-stage characteristics have also been exploited. In process engineering, Gong and You (2018) introduce a multi-objective two-stage adaptive robust model allowing to optimize resilience and economic objectives simultaneously. However, no design selection is provided. Recently, Gabrielli et al. (2019) have proposed an approach in which Pareto-efficient designs are assessed by performance indicators measuring the robustness and the cost optimality. The final selection of one design depends on the target levels which need to be provided by the decision maker.

Until now, taking uncertainty via robust optimization into account when automatically selecting one single design has not been investigated.

CHAPTER 3

Contribution and structure of this thesis

In this chapter, we state the contribution and introduce the elements of the proposed framework of this thesis (Section 3.1). Furthermore, we provide an structural overview of the thesis (Section 3.2).

3.1 Contribution of this thesis

The thesis answers the major open research question:

> *How to select one single sustainable energy system design which ensures security of energy supply despite uncertain component availability and uncertain energy demands?*

The research question is motivated by three gaps in literature which have been revealed in the review in Chapter 2. Hence, the major open research question decomposes into three parts:

I How to regard failure of components in the design optimization of DESS providing multiple energy types?

II How to identify robust system designs which ensure sufficient energy supply while being cost efficient and well-performing regarding sustainability?

III How to select one design when considering multiple criteria by taking advantage of the two-stage character of energy systems?

To answer each part, we propose the following mathematical optimization approaches.

Part I: Reliability

We propose the $(n-1)$-*reliability approach* to identify optimal reliable designs for energy systems (Chapter 5). The approach guarantees energy supply during the failure of one component at any time and is independent of probabilities and the selection of scenarios. Hence, $(n-1)$-reliability also allows for time-flexible maintenance of components. For problems with high computational effort, we propose the inexact but computationally efficient $(n-1^{max})$-*reliability approach* which also guarantees energy supply but allows overproduction (Chapter 5). The $(n-1^{max})$-reliability approach allows easy combination with further approaches as only additional constraints need to be added to the original problem formulation. The reliability approaches allow for uncertainty of energy supply and, thus, avoid a shortage of energy supply due to failure of components which might lead to unexpected high costs or environmental damage. Both approaches not only identify reliable designs ensuring energy supply but designs which are both reliable *and* cost-efficient.

Part II: Robustness

The ***T****wo-stage* ***Robus****tness* ***T****rade-off (TRusT) approach* considers the trade-off between expected costs in the nominal scenario and costs in the worst case while guaranteeing sufficient energy supply (Section 6.1). Thereby, the approach identifies balanced robust energy supply systems which are cost-efficient in both the everyday business and the worst case. The TRusT approach regards uncertainty of input parameters. By considering uncertainty of input parameters during optimization, the designed system is able to cover higher occurring demands than expected in the nominal scenario. Furthermore, the identified design is also robust against volatile energy prices. Thus, the TRusT approach is an appropriate tool to design cost-efficient and secure energy systems.

We apply *minmax* ***Ro****bust* ***M****ulti-objective* ***O****ptimization (RoMO)* (Ehrgott et al., 2014) for designing sustainable energy supply systems. The proposed MILP formulation allows to easily identify robust sustainable designs guaranteeing sufficient energy supply (Section 6.2). To identify sustainable design options, multi-objective optimization is applied (Section 2.3). RoMO combines uncertainty of input parameters with the optimization of multiple criteria. Thus, the identified design options are both, robust and well-performing in aspects of sustainability.

In the context of robust optimization, we always refer to *minmax* robust optimization, and hence, we use the terms *robust* and *minmax robust* synonymously.

Part III: Sustainable design selection

Multi-objective optimization usually leaves several design options to the decision maker to select from (Hennen et al., 2017). The ***flexible here-and-now decision (flex-hand) approach*** automatically selects the best design based on the Pareto front achieved by multi-objective optimization. We refer to the *best* design as the design with highest adaptability to the Pareto front in operation regarding the considered criteria for sustainability. To automatically select the design with highest adaptability, the approach minimizes the distance of the Pareto front based on one fixed design to the Pareto front allowing multiple designs. As a result, the selected design is highly flexible in operation. To not exclude design options a priori, the approach considers designs even beyond the pre-calculated efficient design options on the Pareto front.

The be-rebust framework: Framework to select the best reliable and robust sustainable design

The thesis combines the approaches for reliability, robustness, and sustainable design selection (Fig. 3.1). The combination of these approaches into the be-rebust framework enables to select one single sustainable energy system design which ensures security of energy supply despite uncertain component availability and uncertain energy demands. Thus, the proposed framework answers the major open research question stated in the beginning of this section.

3.2 Structure of this thesis

Before expounding the framework, the basic model for optimal design of DESS and the considered real-world case study of this thesis are introduced (Section 4). Chapter 5 introduces the reliability approaches. The robustness approaches, TRusT and RoMO, are expounded in Section 6.1 and Section 6.2, respectively. The TRusT approach and RoMO are combined in Section 6.3. In Chapter 7, the flex-hand approach for the selection of one single sustainable design is described. The approaches are unified to the be-rebust framework in Chapter 8. In Section 8.1.1, the unification scheme is expounded: In a first step, reliability and robustness

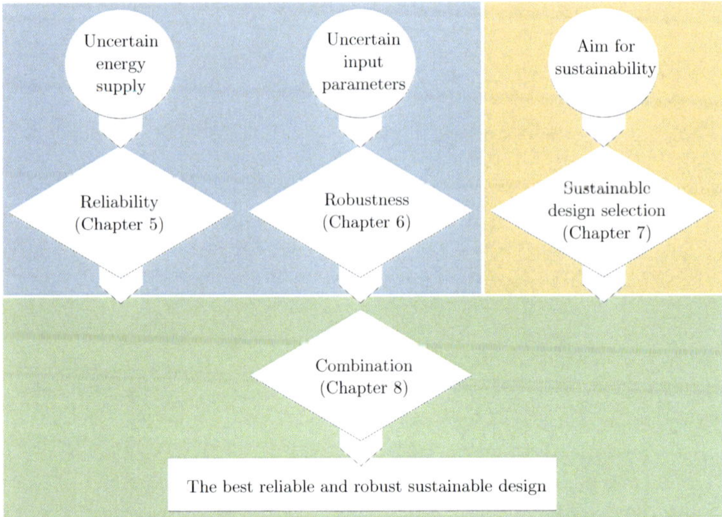

Figure 3.1: Overview of the elements of the be-rebust framework of DESS (framework to select the best **re**liable and ro**bu**st **s**us**t**ainable design – green area); the approaches integrating uncertainties (blue area) are combined with the approach integrating sustainable design selection (yellow area)

are combined (Section 8.1.2). In a second step, the sustainable design selection is adapted to regard for reliability and robustness and is applied to select the best design (Section 8.1.3). Chapters 5-8 all consist of an introduction part of the approach followed by results of a real-world case study. Conclusions for each approach are also included in the respective chapters. For an overview of the proposed approaches see Fig. 3.1. Chapter 9 summarizes and concludes the thesis and provides perspectives of future research.

Chapter 4

Optimal design of distributed energy supply systems

In Chapter 4, first, the model formulation of a DESS is provided. In the second part, the real-world case study is introduced which is considered for all computational calculations of this thesis.

4.1 The optimization model

The considered model of DESS has been published by Voll et al. (2013). The model is formulated as an MILP which is often employed for the synthesis of DESS in literature (Chapter 2). The considered energy conversion components comprise typical components such as absorption chillers $\mathcal{AC}$, compression chillers $\mathcal{CC}$, boilers $\mathcal{B}$, and combined heat and power engines $\mathcal{CHP}$. However, an extension to renewable energies can easily be performed (Baumgärtner et al., 2019b). The optimization problem is formulated as an MILP and identifies an optimal design (structure of DESS and sizing of components) as well as the optimal operation of the designed DESS. In the model proposed by Voll et al. (2013), no failure of components nor parameter uncertainties are taken into account; and thus, security of energy supply can not be ensured. In the following, we call this model *nominal* to clearly distinguish between the model without considering any uncertainty, neither uncertainty of energy supply nor uncertainty of input parameters.

4.1.1 Objective functions

The brief review in Section 2.3 shows that most frequently aggregated economic criteria and environmental criteria are considered in energy system optimization. Thus, we focus on the trade-off between aggregated costs and ecology in this work. To present the methodological approaches of this thesis, we exemplarily choose the often employed total annualized costs TAC as economic criterion, and, the global warming impact GWI as environmental criterion. However, other environmental criteria as employed in life cycle assessment, e. g., by Huijbregts et al. (2016), could also be integrated in the proposed approaches.

The total annualized costs TAC are defined by

$$
\begin{aligned}
& TAC\left(\dot{U}, \dot{U}^{el,buy}, \dot{V}^{el,sell}, I\right) \\
&= \sum_{t\in\mathcal{T}} \left[\Delta\tau_t \left(\tilde{p}^{gas} \cdot \sum_{k\in\mathcal{B}\cup\mathcal{CHP}} \dot{U}_{kt} + p^{el,buy} \cdot \dot{U}_t^{el,buy} - p^{el,sell} \cdot \dot{V}_t^{el,sell}\right)\right] \\
&\quad + \sum_{k\in\mathcal{K}} \left(\frac{1}{PVF} + p_k^m\right) \cdot I_k ,
\end{aligned} \tag{4.1}
$$

where k represents a component in the set of all components $\mathcal{K} = \mathcal{B}\cup\mathcal{CHP}\cup\mathcal{AC}\cup\mathcal{CC}$ which might be installed. For each time step $t \in \mathcal{T}$, $\Delta\tau_t$ represents its length. The corresponding input energy flows of natural gas for boilers $\mathcal{B}$ and combined heat and power engines $\mathcal{CHP}$ are denoted by $\sum_{k\in\mathcal{B}\cup\mathcal{CHP}} \dot{U}_{kt}$. The system's input and the output energy flow of electricity are denoted by $\dot{U}_t^{el,buy}$ and $\dot{V}_t^{el,sell}$, respectively. The annual maintenance costs of unit k are expressed as a share of the investment costs by the factor p_k^m. For annualizing the investment costs I_k, we use the present value factor (Broverman, 2010)

$$
PVF = \frac{(r+1)^{\vartheta} - 1}{(r+1)^{\vartheta} \cdot r}
$$

with an interest rate r and a time horizon ϑ. All newly installed components have capacity-dependent investment costs I_k which are piece-wise linearized (Appendix A.1). In the case study, the first stage consists of the design variables (see Appendix A.2), besides the investment costs I_k for each installed component k. The second-stage variables comprise time-dependent operational variables $\dot{U}_{kt}$, $\dot{U}_t^{el,buy}$ and $\dot{V}_t^{el,sell}$. Instead of the total annualized costs TAC, any other economic objective function can be chosen, e. g., investment costs (Appendix C.2).

The global warming impact GWI is given by

$$\begin{aligned} & GWI\left(\dot{U}, \dot{U}^{el,buy}, \dot{V}^{el,sell}\right) \\ & = \sum_{t\in\mathcal{T}} \Delta\tau_t \left[\sum_{k\in\mathcal{B}\cup\mathcal{CHP}} \dot{U}_{kt} \cdot GWI^{gas} + \left(\dot{U}_t^{el,buy} - \dot{V}_t^{el,sell}\right) \cdot GWI^{el} \right], \end{aligned} \tag{4.2}$$

where GWI^{gas} and GWI^{el} define the specific global warming impact of gas and of the electricity mix of the grid, respectively. When selling electricity to the grid, a credit for global warming impact is given, following the idea of avoided burden (Baumann and Tillman, 2014). In Eq. (4.2), the global warming impact GWI depends only implicitly on design variables x^f. A direct influence would be given if the global warming impact induced by the manufacturing of the components was taken into account. Since the global warming impact of the operation has usually a significantly higher impact (Guillén-Gosálbez, 2011), we neglect this dependency. However, this simplification does not affect any of the presented approaches of this thesis. Hence, all presented approaches are also applicable in case of a highly detailed formulation of the global warming impact GWI.

4.1.2 Energy balances

Operation of components is possible from the installed nominal thermal power down to a technology-specific minimal part load. For the nominal problem, the energy balances are given by:

$$\sum_{k\in\mathcal{K}} \dot{V}_{kt} = \dot{E}_t \qquad \forall t \in \mathcal{T}. \tag{4.3}$$

$\dot{V}_{kt}$ represents the supplied energy flow of component k in time step t. The energy demands in time step t are given by $\dot{E}_t$. The demands consist of heating demands $\dot{E}_t^{h,tot}$, cooling demands $\dot{E}_t^c$, and electricity demands $\dot{E}_t^{el}$. For heating, the energy demand $\dot{E}_t^{h,tot}$ comprises not only the heating requirements on the industrial site $\dot{E}_t^h$, but also the necessary energy $\sum_{k\in\mathcal{AC}} \frac{\dot{V}_{kt}}{\eta_k}$ for running the absorption chillers $k \in \mathcal{AC}$ with coefficient of performance η_k:

$$\dot{E}_t^{h,tot} = \dot{E}_t^h + \sum_{k\in\mathcal{AC}} \frac{\dot{V}_{kt}}{\eta_k} \qquad \forall t \in \mathcal{T}. \tag{4.4}$$

We substitute the ration $\frac{\dot{V}_{kt}}{\eta_k}$ by the input energy flow $\dot{U}_{kt}$ and yield the following formulation for heating and cooling circuits:

$$\begin{aligned}
&\sum_{k \in \mathcal{B} \cup \mathcal{CHP}} \dot{V}_{kt} - \sum_{k \in \mathcal{AC}} \dot{U}_{kt} = \dot{E}_t^h && \forall t \in \mathcal{T} \\
&\sum_{k \in \mathcal{AC} \cup \mathcal{CC}} \dot{V}_{kt} = \dot{E}_t^c && \forall t \in \mathcal{T} \\
&\sum_{k \in \mathcal{CHP}} \dot{V}_{kt}^{el} - \sum_{k \in \mathcal{CC}} \dot{U}_{kt}^{el} + \dot{U}_t^{el,buy} - \dot{V}_t^{el,sell} = \dot{E}_t^{el} && \forall t \in \mathcal{T}\,.
\end{aligned} \tag{4.5}$$

The full problem formulation of the model for DESS optimization is presented in Appendix A.2.

Even though the optimal synthesis of DESS is NP-hard, as proofed by Goderbauer et al. (2019), common available solvers are often able to obtain solutions in reasonable time. For large-scale problem formulations, special tailor-made solution methods might be necessary (Bahl et al., 2018a; Baumgärtner et al., 2019a).

4.2 Uncertainty of input parameters

In this thesis, we include uncertainty in DESS design. In general, input parameters of the optimization problem are not known with perfect foresight. However, the original single-objective optimization problem by Voll et al. (2013) neglects the uncertainty of input parameters. Since neglecting uncertainties leads to sub-optimal and even infeasible solutions, we take uncertainties into account when designing a sustainable energy system. In this thesis, we consider three types of typically uncertain parameters:

- the energy prices,
- the specific global warming impact of the electricity provided by the grid, and
- the energy demands on site.

The proposed robustness approaches (Section 6.1 and Section 6.2) could also be extended to include other uncertainties in the input parameters, e. g., efficiencies of the components.

The considered uncertainties affect the energy balances as well as the global warming impact GWI and the total annualized costs TAC.

Uncertainty is expressed by a tilde sign above the uncertain parameters. Every set of uncertain parameters represents one *scenario* ξ. Usually, demands and energy tariffs as well as the specific global warming impacts fluctuate within some range. For this reason, *interval-based uncertainty* is considered. Thereby, the uncertainty of each parameter is expressed by an upper and a lower bound of the variation around the nominal value (Ben-Tal and Nemirovski, 1999). Additionally, we restrict demands to positive values. Thus, the *uncertainty set* $\mathcal{U} \subseteq \mathbb{R}^u$ containing all potentially occurring scenarios ξ is given by

$$\begin{aligned}
\mathcal{U} = \Bigg\{ \Bigg(& \tilde{p}^{gas}, \tilde{p}^{el,sell}, \tilde{p}^{el,buy}, \widetilde{GWI}^{el}, \left(\tilde{\dot{E}}_t^{h}\right)_{t \in \mathcal{T}}, \left(\tilde{\dot{E}}_t^{c}\right)_{t \in \mathcal{T}}, \left(\tilde{\dot{E}}_t^{el}\right)_{t \in \mathcal{T}} \Bigg) \Bigg| \\
& \tilde{p}^{gas} = \hat{p}^{gas}(1 + pg), \ pg \in [-\delta^{pg}, \delta^{pg}]; \\
& \tilde{p}^{el,sell} = \hat{p}^{el,sell}(1 + pe), \\
& \tilde{p}^{el,buy} = \hat{p}^{el,buy}(1 + pe), \ pe \in [-\delta^{pe}, \delta^{pe}]; \\
& \widetilde{GWI}^{el} \in \left[\widehat{GWI}^{el} - \underline{\delta}^{ge}, \widehat{GWI}^{el} + \overline{\delta}^{ge} \right], \\
& \tilde{\dot{E}}_t^{h} \in \left[\max\left\{0, \hat{\dot{E}}_t^{h} - \delta_t^{\dot{E}h}\right\}, \hat{\dot{E}}_t^{h} + \delta_t^{\dot{E}h} \right], \\
& \tilde{\dot{E}}_t^{c} \in \left[\max\left\{0, \hat{\dot{E}}_t^{c} - \delta_t^{\dot{E}c}\right\}, \hat{\dot{E}}_t^{c} + \delta_t^{\dot{E}c} \right], \\
& \tilde{\dot{E}}_t^{el} \in \left[\max\left\{0, \hat{\dot{E}}_t^{el} - \delta_t^{\dot{E}e}\right\}, \hat{\dot{E}}_t^{el} + \delta_t^{\dot{E}e} \right], \\
& \qquad \delta^{pg}, \delta^{pe}, \underline{\delta}^{ge}, \overline{\delta}^{ge}, \delta_t^{\dot{E}h}, \delta_t^{\dot{E}c}, \delta_t^{\dot{E}e} \geq 0, t \in \mathcal{T} \Bigg\},
\end{aligned} \tag{4.6}$$

where $\hat{\xi} = \left(\hat{p}^{gas}, \hat{p}^{el,sell}, \hat{p}^{el,buy}, \left(\hat{\dot{E}}_t^{h}\right)_{t \in \mathcal{T}}, \left(\hat{\dot{E}}_t^{c}\right)_{t \in \mathcal{T}}, \left(\hat{\dot{E}}_t^{el}\right)_{t \in \mathcal{T}} \right)$ defines the nominal scenario, which corresponds to the parameters which would have been employed in the nominal problem with perfect foresight. The maximal deviation from the nominal values is given by δ with corresponding indices, e. g., we use $\delta_t^{\dot{E}h}$ for the maximal deviation of the heating demands in time step t. For electricity prices, the variation of the prices for purchasing $p^{el,buy}$ and feeding in electricity $p^{el,sell}$ are coupled by the parameter pe because their price levels are correlated. $\widehat{GWI}^{el} - \underline{\delta}^{ge} = \underline{GWI}^{el}$ and $\widehat{GWI}^{el} + \overline{\delta}^{ge} = \overline{GWI}^{el}$ denote the lower and upper bound of the uncertain specific global warming impact $\widetilde{GWI}^{el}$, respectively. Since all uncertain values can take any value within an interval, the uncertainty set $\mathcal{U}$ comprises an infinite number of scenarios.

4.3 Case study: Description of the real-world industrial site

In the comprehensive case study, an industrial park with heating, cooling, and electricity demands is considered (Voll et al., 2013). The industrial park is divided into two areas: *Site A* and *Site B*. Both sites share a common heating system, whereas each site has its own cooling system. The existing energy system on Site A already comprises 2 boilers, 1 combined heat and power engine, and 2 compression chillers. However, there is a new cooling demand on Site B. Thus, a retrofit of the energy system is necessary.

The heating as well as the cooling demands and their uncertainties are shown in Fig. 4.1 (and Tab. A.1). The uncertainties considered in the case study are deduced from real data of previous years.

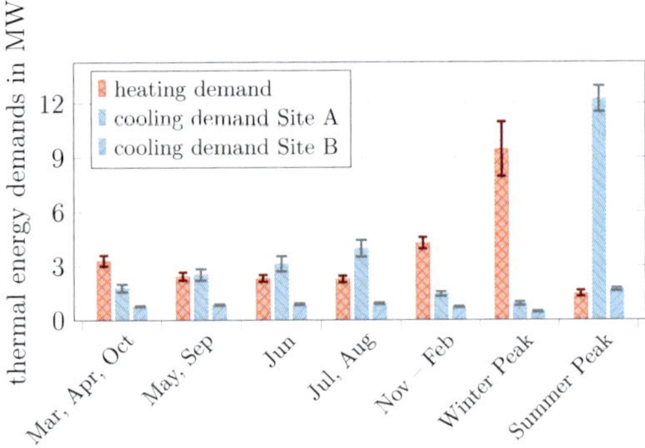

Figure 4.1: Heating and cooling demands on Site A and Site B of the industrial park in the case study with uncertainties indicated by error bars.

The load cases represent aggregated months with similar load profiles. In the nominal problem, the DESS must be able to cover both the aggregated demands *and* the peak demands to ensure sufficient capacity of the energy supply system. Peak loads are always considered, independent of taking uncertainties into account or not. The total deviation between minimal and maximal uncertain cooling demand is 19.4 % of the nominal value which corresponds to 5.2 GWh/a. For the heating demands, the deviation is 16.2 % corresponding to 4.6 GWh/a. The electricity demands range from 34.6 GWh/a to 64.9 GWh/a.

The demands shall be covered simultaneously using the synergy of trigeneration systems. The real-world example includes typical characteristics of DESS and incorporates boilers $\mathcal{B}$, combined heat and power engines $\mathcal{CHP}$, absorption chillers $\mathcal{AC}$, and compression chillers $\mathcal{CC}$ (Section 2.1). For the components operated in part load, we assume constant efficiencies/coefficients of performance. However, minimal part-load power of components is taken into account based on equipment data sheets (Voll et al., 2013): The minimal part load of combined heat and power engines is 50 % of the installed thermal power. All remaining technologies can operate between 20 % of the installed thermal power and full load. Further values for specification of components can be found in Voll (2013).

In the nominal scenario, gas can be purchased at a price of $\widehat{p}^{gas} = 5\,\text{ct/kWh}$. Electricity can be purchased and fed into the grid at prices of $\widehat{p}^{el,buy} = 16\,\text{ct/kWh}$ and $\widehat{p}^{el,sell} = 10\,\text{ct/kWh}$, respectively. The prices are uncertain lying inside a range of $\delta^{pg} = 40\,\%$ for gas and for electricity inside a range of $\delta^{pe} = 46\,\%$ of the nominal values. The uncertainties are deduced from historical data of the EEX spot market. We consider a time horizon of $\vartheta = 4$ years and an interest rate of $r = 8\,\%$.

Since the future electricity mix is uncertain, the specific global warming impact of the electricity GWI^{el} provided by the grid is uncertain. In the case study, we use values for the year 2020. When planning sustainable energy supply systems, future trends have to be considered. There are various forecasts for the electricity mix of Germany. In the case study, we employ 3 forecasts representing a nominal scenario as well as a minimal and a maximal scenario, as described in the following. From the scenarios of electricity mix, we derive corresponding values for the specific global warming impact of the electricity mix GWI^{el} purchased from the grid using the software GaBi (thinkstep, 2016). The minimal value corresponds to a scenario aspiring an 80 % decrease of CO_2 emissions until 2050 (Nitsch et al., 2012). The aim of decreasing the CO_2 emissions by 80 % is ambitious. Hence, we choose this scenario as lower bound of the specific global warming impact with $\underline{GWI}^{el} = 430\,\text{g}_{\text{CO}_2\text{-eq.}}/\text{kWh}$ for 2020. The nominal specific global warming impact of the electricity mix $\widehat{GWI}^{el} = 561\,\text{g}_{\text{CO}_2\text{-eq.}}/\text{kWh}$ is deduced from a forecast of the German Energy Agency (dena) (Kohler et al., 2010). In 2016, the value corresponded to about $610\,\text{g}_{\text{CO}_2\text{-eq.}}/\text{kWh}$ (thinkstep, 2016). Since the trend of the specific global warming impact is decreasing based on current international policies (Birol et al., 2015), we select $\overline{GWI}^{el} = 610\,\text{g}_{\text{CO}_2\text{-eq.}}/\text{kWh}$ as upper bound of the uncertain specific global warming impact $\widetilde{GWI}^{el}$. The global warming impact of the natural gas GWI^{gas} with a value of 244 $g_{CO_2\text{-eq.}}$/kWh is assumed to be certain.

The introduced case study is employed to evaluate the proposed approaches (Chapter 5-7) and the be-rebust framework (Chapter 8). Additional information is provided at the beginning of each analysis in the corresponding chapters. For calculations, the automated superstructure-generation approach from Voll et al. (2013) is applied. The approach successively increases the number of components of each technology to build up a superstructure which contains the optimal design but is not oversized. The largest possible superstructures resulting from this successive approach contains a predefined maximum number of boilers, combined heat and power engines, compression chillers, and absorption chillers. In the case study, the maximal number of components per technology is set to 10.

Chapter 5

Reliability of energy supply

In this chapter, approaches are presented *ensuring* reliability in the design of DESS during the failure of 1 component. Taking uncertainty of energy supply into account is mandatory for the design of secure DESS. The approaches proposed in this chapter introduces uncertainty of energy supply into the framework (Fig. 5.1).

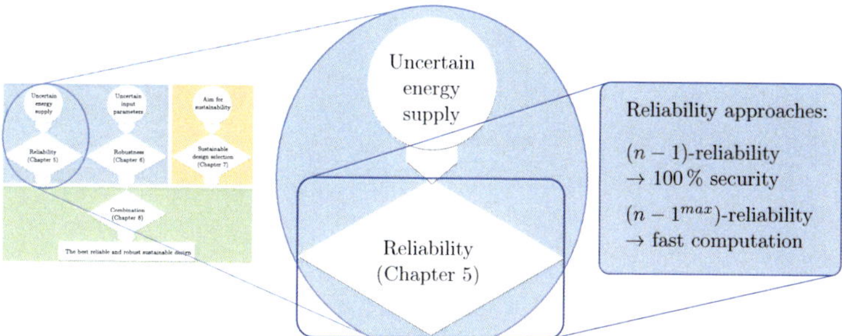

Figure 5.1: Overview Chapter 5 – $(n-1)$-reliability and $(n-1^{max})$-reliability for reliability of energy supply

---

Major parts of this chapter are reproduced by permission of Elsevier from:

> Hollermann, D. E., Hoffrogge, D. F., Mayer, F., Hennen, M., Bardow, A. (2019b) Optimal $(n-1)$-reliable design of distributed energy supply systems. *Computers & Chemical Engineering*, 121:317–326

Contribution report: Principal author, idea of approaches, implementation of the $(n-1)$-reliability approach, analysis of results, writing the draft

The $(n-1)$-reliability approach follows the idea of approaches for reliable power transmission networks (Section 2.4.1). Power transmission network must provide full operation with only $n-1$ instead of n available components. However, for DESS optimization, additional types of final energy need to be considered involving coupling of energy supply circuits, e. g., of heating and cooling circuits. Hence, our $(n-1)$-reliable approach extends existing approaches for network planning of power system engineering. Moreover, the obtained $(n-1)$-reliable design is independent of selecting failure scenarios and independent of failure probabilities. Instead, we consider a possible failure of 1 arbitrary component at any time. The identified reliable design is thus able to cover all energy demands exactly during failure or also during maintenance of 1 component. For problems with high computational effort, we propose an inexact but computationally efficient approach called $(n-1^{max})$-reliability. In this alternative approach, we ensure sufficient energy supply during the failure of any component while allowing overproduction. Overproduction describes an operational state in which more energy is supplied than demanded. Allowing overproduction reduces the optimization problem to the analysis of failure of the largest component and is therefore called $(n-1^{max})$-reliability. The reduced computational effort allows coupling reliability with other optimization approaches.

The $(n-1)$-reliability approach can be directly applied to optimize all types of energy systems. In contrast, the $(n-1^{max})$-reliability approach needs further adaption the optimized system (Section 5.1.2).

Chapter 5 is structured as follows: In Section 5.1, we introduce the $(n-1)$-reliable approach and the $(n-1^{max})$-reliable approach. The approaches are applied to the real-world case study in Section 5.2. We summarize and conclude Chapter 5 in Section 5.3.

5.1 Reliable design of distributed energy supply systems

To guarantee secure energy supply, we propose two reliability approaches. We present the approaches for a tri-generation system providing electricity, heating, and cooling energy. In our context, we assume that electricity demands can always be covered by the electricity grid; thus, we focus on covering heating and cooling demands during failure of components. A motivating example related to everyday life is provided in Appendix B.1.

5.1.1 $(n-1)$-reliability: An exact approach for the optimal reliable design

The idea of $(n-1)$-reliability is that the designed energy system can compensate the failure of any component out of n components at any time – but only 1 component at the same time. Thus, a failure of 1 component during maintenance of another might still lead to a lack of energy supply. In other words, we aim for an $(n-1)$-out-of-n system (Birnbaum, 1968) in which $n-1$ components out of n need to be fully available to guarantee a 100 percent reliable system. To model $(n-1)$ reliability, we consider all possible scenarios for a failure of 1 component. To implement $(n-1)$-reliability, we introduce additional energy balances for heating and cooling circuits describing the assumption that any component might fail at any time and cannot provide energy:

$$\sum_{k\in\mathcal{K}^h\setminus\{k'\}} \dot{V}_{kt}^{k'} = \dot{E}_t^h + \sum_{k\in\mathcal{AC}\setminus\{k'\}} \frac{\dot{V}_{kt}^{k'}}{\eta_k} \qquad \forall t\in\mathcal{T} \quad \forall k'\in\mathcal{K} \tag{5.1}$$

$$\sum_{k\in\mathcal{K}^c\setminus\{k'\}} \dot{V}_{kt}^{k'} = \dot{E}_t^c \qquad \forall t\in\mathcal{T} \quad \forall k'\in\mathcal{K}. \tag{5.2}$$

The set of components $\mathcal{K}$ is divided into heating $\mathcal{K}^h$ and cooling components $\mathcal{K}^c$. Every possibly failing component $k'\in\mathcal{K}$ leads to a new set of energy balances. To formulate these additional energy balances, we introduce new operation variables $\dot{V}_{kt}^{k'}$ which specify the adapted output of component k in time step t if component k' fails. The new operation variables are determined such that the demands can still be covered exactly with the remaining components $\mathcal{K}\setminus\{k'\}$.

Since the additional variables $\dot{V}_{kt}^{k'}$ only appear in the constraints and not in the objective function, they only impact on the design of the DESS, i. e., the selected components and their sizing. The objective function is unchanged which allows replacing the criterion of the optimization easily.

The problem of $(n-1)$-reliability can also be formulated using robust optimization. For this purpose, adjustable robustness (Ben-Tal et al., 2004) can be employed to describe the adaption of DESS to a failing component: The remaining components adjust their amount of provided energy on the second stage such that the demands can still be fulfilled:

$$\sum_{k\in\mathcal{K}^h} \dot{V}_{kt}(\zeta_{kt}) - \zeta_{kt}\dot{V}_{kt}(\zeta_{kt}) = \dot{E}_t^h + \sum_{k\in\mathcal{AC}} \frac{\left(\dot{V}_{kt}(\zeta_{kt}) - \zeta_{kt}\dot{V}_{kt}(\zeta_{kt})\right)}{\eta_k} \qquad \forall t \in \mathcal{T}, \zeta_{kt} \in \mathcal{U}^{rely} \tag{5.3}$$

$$\sum_{k\in\mathcal{K}^c} \dot{V}_{kt}(\zeta_{kt}) - \zeta_{kt}\dot{V}_{kt}(\zeta_{kt}) = \dot{E}_t^c \qquad \forall t \in \mathcal{T}, \zeta_{kt} \in \mathcal{U}^{rely}. \tag{5.4}$$

The additional operation variables $\dot{V}_{kt}(\zeta_{kt})$ depend on the scenario ζ_{kt} with $\zeta_{kt} = 1$ describing the failure of component k in time step t. The corresponding uncertainty set is given by

$$\mathcal{U}^{rely} := \left\{ \left(\zeta'_{kt}\right)_{k\in\mathcal{K},t\in\mathcal{T}} \in \{0,1\}^{|\mathcal{K}|\times|\mathcal{T}|} : \sum_{k\in\mathcal{K}} \zeta'_{kt} = 1, \forall t \in \mathcal{T} \right\}. \tag{5.5}$$

The sum $\sum_{k\in\mathcal{K}} \zeta'_{kt} = 1$ ensures that only 1 component fails at the same time.

$(n-1)$-reliability can be extended such that the failure of more than 1 component is taken into account as done for electricity grids by Street et al. (2014). For DESS, using the notation of Eq.s (5.1) and (5.2), the following equations need to be added to the nominal problem:

$$\sum_{k\in\mathcal{K}^h\setminus\{\mathcal{K}'\}} \dot{V}_{kt}^{\mathcal{K}'} = \dot{E}_t^h + \sum_{k\in\mathcal{AC}\setminus\{\mathcal{K}'\}} \frac{\dot{V}_{kt}^{\mathcal{K}'}}{\eta_k} \qquad \forall t \in \mathcal{T} \quad \forall \mathcal{K}' \in \mathcal{P}(\mathcal{K}) \tag{5.6}$$

$$\sum_{k\in\mathcal{K}^c\setminus\{\mathcal{K}'\}} \dot{V}_{kt}^{\mathcal{K}'} = \dot{E}_t^c \qquad \forall t \in \mathcal{T} \quad \forall \mathcal{K}' \in \mathcal{P}(\mathcal{K}). \tag{5.7}$$

where $\mathcal{K}'$ is a subset of the power set $\mathcal{P}(\mathcal{K})$ including all combinations of possibly failing heating and cooling components. The cardinality of components failing at the same time can be limited, i. e., $|\mathcal{K}'| \leq K$. This formulation guarantees $(n-K)$-reliability. $(n-K)$-reliability leads to an exponential increase of the number of additional equations proportional to $\sum_{i=1}^{K} \left(\binom{|\mathcal{K}|}{i} \cdot |\mathcal{T}|\right)$ and thus of computational time. For DESS, $(n-1)$-reliability, i. e., $K = 1$, is the most basic and commonly applied method of $(n-K)$-reliability in engineering practice. Eq.s (5.6) and (5.7) are equivalent to Eq.s (5.1) and (5.2) when the failure of only 1 component is considered. Considering the failure of 1 arbitrary component is often sufficient (Aguilar et al., 2008). Thus, in this thesis, the focus is set on $(n-1)$-reliable design of DESS. Still, even for this case study, the additional energy balances lead to many additional equality constraints and the problem size scales up polynomially in two variables: The number of additional constraints increases proportionally with the number of potentially installed components $|\mathcal{K}|$ times the number of considered time steps $|\mathcal{T}|$.

Hence, employing $(n-1)$-reliability leads to an increased problem size which might result in high computational times. Therefore, in the following section, we propose an alternative, computationally fast approach while still ensuring sufficient energy supply.

5.1.2 $(n-1^{max})$-reliability: An inexact approach for the optimal design ensuring sufficient energy supply

The idea of $(n-1^{max})$-reliability is to be able to supply at least any required heating and cooling demand during a failure of any single component at any time while allowing for overproduction. If overproduction is allowed, it is sufficient to consider only the failure of the largest cooling or heating component to cover the failure of any component: A larger component can always replace a smaller component. The only limitation could arise from minimal part-load constraints for the larger component. In the proposed approach, this limitation is overcome by allowing overproduction during failure. Any excess energy produced during overproduction has to be wasted, i. e., released to the environment. Alternatively, a component could be operated below the specified minimal part-load power.

To be able to supply at least all heating demands $\dot{E}_t^{h,tot}$ and cooling demands $\dot{E}_t^c$ during failure of *1 arbitrary* component, we ensure sufficient installed thermal power during the failure of the *largest* component. Hence, we force the sum of the nominal thermal power of the all components except the largest component to be sufficient to cover all occurring demands:

$$\sum_{k\in\mathcal{K}^{h/c}} \dot{V}_k^N - \dot{V}_{\mathcal{K}^{h/c}}^{max} \geq \dot{E}_t^{h,tot/c} \qquad \forall t \in \mathcal{T}\,. \tag{5.8}$$

Again, $\mathcal{K}^h$ denotes heating components and $\mathcal{K}^c$ cooling components. $\dot{V}_k^N$ represents the installed thermal power of component k and $\dot{V}_{\mathcal{K}^{h/c}}^{max}$ the maximal deficit in heating/cooling supply induced due to failure of 1 component. The maximal deficit $\dot{V}_{\mathcal{K}^{h/c}}^{max}$ is defined as the maximal installed heating/cooling power of 1 single component:

$$\dot{V}_{\mathcal{K}^{h/c}}^{max} := \max \left\{ \dot{V}_k^N \,\middle|\, k \in \mathcal{K}^{h/c} \right\} \tag{5.9}$$

which can be reformulated by

$$\dot{V}_{\mathcal{K}^{h/c}}^{max} \geq \dot{V}_k^N \qquad \forall k \in \mathcal{K}^{h/c}. \tag{5.10}$$

Since the heating demands $\dot{E}_t^{h,tot}$ also depend on the operation of the absorption chillers (Eq. (4.4)), a failure in a cooling circuit might also affect the heating circuit: A failing compression chiller could be replaced by absorption chillers. The replacing absorption chillers increase the heating demand. This dependency needs to be taken into account for reliability:

$$\sum_{k \in \mathcal{K}^h} \dot{V}_k^N \geq \dot{E}_t^h + \sum_{k \in \mathcal{AC}} \frac{\dot{V}_{kt}}{\eta_k} + \dot{E}_{\mathcal{AC}}^h \qquad \forall t \in \mathcal{T}. \tag{5.11}$$

The installed heating power $\sum_{k \in \mathcal{K}^h} \dot{V}_k^N$ needs to cover at least the current total heating demand of each time step t, i. e., the sum of $\dot{E}_t^h$ and $\sum_{k \in \mathcal{AC}} \frac{\dot{V}_{kt}}{\eta_k}$. $\dot{E}_{\mathcal{AC}}^h$ adds the maximal heating supply due to failure of a cooling component:

$$\dot{E}_{\mathcal{AC}}^h = \min \left\{ \max_{k \in \mathcal{CC}} \frac{\dot{V}_k^N}{\eta_{\mathcal{AC}}^{min}} \; ; \; \sum_{k \in \mathcal{AC}} \frac{\dot{V}_k^N}{\eta_k} \right\}. \tag{5.12}$$

The first term, $\frac{\dot{V}_k^N}{\eta_{\mathcal{AC}}^{min}}$, describes the heating energy needed to replace the largest compression chiller by absorption chillers considering their worst coefficient of performance $\eta_{\mathcal{AC}}^{min}$. The second term, $\sum_{k \in \mathcal{AC}} \frac{\dot{V}_k^N}{\eta_k}$ corresponds to the maximal heating energy required by the absorption chillers. We propose the employed MILP reformulation of Eq. (5.12) in Appendix B.2.1. An even tighter but more complex approximation of the maximal additional heating demand is additionally provided in Appendix B.2.2.

A failing heating component $k \in \mathcal{K}^h$ might also affect a cooling circuit, since absorption chillers depend on the provided heating energy. However, this coupling of circuits is already regarded by employing Eq. (5.8). Due to the specific interdependence of circuits, the $(n - 1^{max})$-reliability approach is not directly applicable to any other energy system optimization problem but has to be adapted to the specific system considered.

To model the ability to cover at least the required demands during the failure of 1 arbitrary component at any time, we only need to add Eq.s (5.8) and (5.12) to the optimization problem. Since again only constraints are added to the nominal problem, applying the $(n-1^{max})$-reliable approach allows changing the objective function as easily as in the $(n-1)$-reliable approach.

The advantage of $(n-1^{max})$-reliability compared to $(n-1)$-reliability is that the number of additional constraints and variables is significantly reduced. The number of additional constraints does not depend on the product of the number of time steps $|\mathcal{T}|$ and components $|\mathcal{K}|$ but only on the sum. Eq. (5.8) involves $2 \cdot |\mathcal{T}|$ additional equations which can be further reduced, since the maximal cooling demand can be determined in a pre-processing step. Since the constraints for calculating the maximal deficit (Eq. (5.10)) only depend on the set of potentially installed components $\mathcal{K} = \mathcal{K}^h \cup \mathcal{K}^c$ but not on the set of time steps $\mathcal{T}$, the number of additional constraints is proportional to the number of potentially installed components $|\mathcal{K}|$. The number of additional equations for introducing Eq. (5.12) is proportional to the number of cooling components $|\mathcal{K}^c|$ (Appendix B.2.1). Considering all additional equations, the problem complexity increases proportionally to the sum of number of time steps and components $|\mathcal{T}| + |\mathcal{K}|$. As a result, the expected increase in computational time compared to the nominal problem is low – at the cost of potential overproduction.

The proposed reliability approaches focus on designing a DESS reliable in steady state. However, dynamic short-term effects such as the starting behavior of components might also lead to shortage of energy supply for a short period. By including storage systems with imposed minimal storage level, such short-term shortage of energy supply can be avoided. A reasonable minimal storage level can be determined such that the level only depends on the starting time and the size of the components without depending on time steps $t \in \mathcal{T}$ (Ihlemann, 2017). In practical applications, pipes of the energy systems have often a large storage capacity (Li et al., 2016) which can be used to cover short-term deficits.

5.2 Results of the case study

In this section, both proposed reliability approaches are applied to a real-world case study of an industrial park (Section 4.3). The problems are implemented in GAMS 24.7.3 (McCarl and Rosenthal, 2016) and CPLEX 12.6.3.0 (IBM Corporation, 2015) is used to solve the problems to machine accuracy on a computer with 3.24 GHz and 64 GB RAM employing 4 threads. All values regarding model scales and computational times relate to the problem with the largest superstructure which is solved applying the automated superstructure-generation approach from Voll et al. (2013).

For optimization, we assume a "green field" without existing energy system components on the industrial site. However, the proposed reliability approaches can also be employed for reliable optimal retrofit of DESS. Expected solutions would comprise existing components with possibly poor efficiencies as spare components to prevent additional investment costs.

To evaluate the reliability approaches introduced in Section 5.1.1 and 5.1.2, we consider a pragmatic approach from engineering practice as benchmark: After identifying an optimal design for the nominal case, the respective largest component of each heating and cooling circuit is installed twice to increase reliability of the DESS while allowing overproduction. We call this heuristic approach *$(2 \times max)$-reliability*. The corresponding design is called *$(2 \times max)$-reliable design* $(x^f)^{(2\times max)}$.

5.2.1 Reliable designs for distributed energy supply systems

In the following, we compare results of the nominal problem to the heuristic $(2 \times max)$-reliable approach, the inexact $(n-1^{max})$-reliable approach (Section 5.1.2), and the exact $(n-1)$-reliable approach (Section 5.1.1). For this purpose, we present the selected components and their sizing (i. e., the optimal design) for the nominal problem and the three reliability approaches (Fig. 5.2).

The $(2 \times max)$-reliable design $(x^f)^{(2\times max)}$ comprises the same components as the nominal design $\widehat{x}^f$ but includes twice the largest component of each circuit, according the idea of the approach. This approach usually leads to high investment costs and, in general, failure of a small component might lead to high overproduction. In the presented case study, the $(2 \times max)$-reliable design $(x^f)^{(2\times max)}$ can compensate the failure of any component and is able to fulfill the $(n-1)$-reliable energy balances (Eq. (5.1) and (5.2)). In general, there could be cases where the $(2 \times max)$-reliable design is not fully reliable: If an absorption chiller is installed twice and the only compression chiller fails, the additionally needed heating supply $\dot{E}^h_{AC}$ might not be covered by the installed components. However, this scenario seems rather fabricated and unlikely in practice as high heating demands and high cooling demands usually do not appear simultaneously.

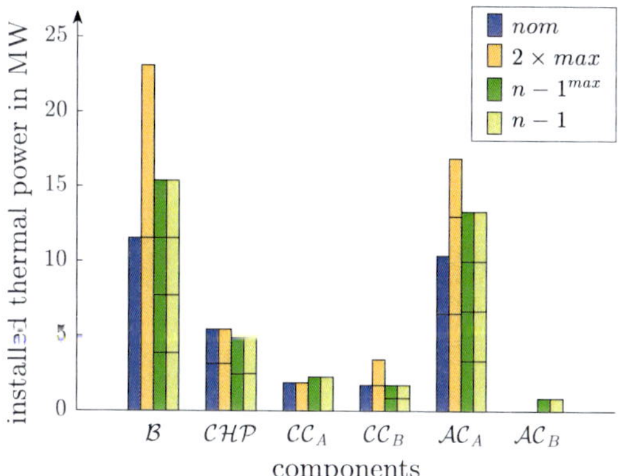

Figure 5.2: Installed components and their installed thermal capacities for each approach: components of nominal (nom) design $\widehat{x}^f$ (without considering failure), of $(2 \times max)$-reliable design $(x^f)^{(2\times max)}$ (doubling the largest component of the nominal design for each circuit), of $(n-1^{max})$-reliable design $(x^f)^{(n-1^{max})}$ (inexact approach allowing failure of any component but allowing overproduction), and of $(n-1)$-reliable design $(x^f)^{(n-1)}$ (exact method allowing failure of any component) for each technology; $\mathcal{B}$ boiler, $\mathcal{CHP}$ combined heat and power engine, $\mathcal{CC}_A$ and $\mathcal{CC}_B$ compression chillers, and $\mathcal{AC}_A$ and $\mathcal{AC}_B$ absorption chillers installed on Site A and Site B, respectively

Employing $(n-1^{max})$-reliability and $(n-1)$-reliability leads to nearly identical reliable designs: For most component types, the designs comprise more and smaller components than the nominal and the $(2 \times max)$-reliable design. As a result, the designs have a higher system flexibility. In general, the $(n-1^{max})$-reliable design $(x^f)^{(n-1^{max})}$ cannot guarantee to cover energy demands exactly and might lead to overproduction if small components fail. Such designs are discussed in Section 5.2.2. In the presented case study, no overproduction occurs during failure, if the $(n-1^{max})$-reliable design $(x^f)^{(n-1^{max})}$ is implemented.

The increase in costs compared to the nominal problem differs for the three reliability approaches (Fig. 5.3).

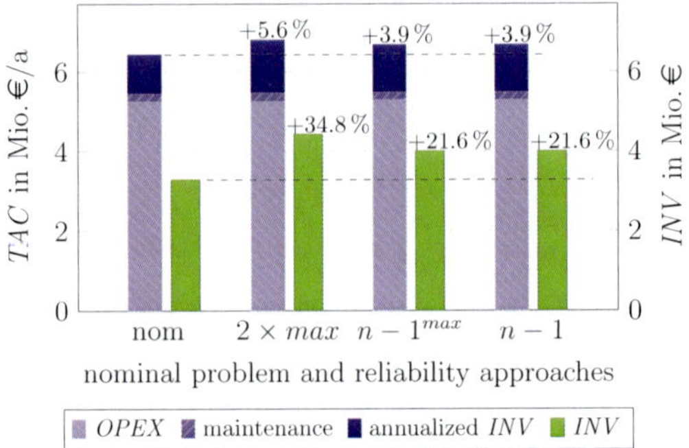

Figure 5.3: Total annualized costs TAC, annualized operational costs $OPEX$, and (annualized) investment costs INV for the designs of nominal problem and the reliability approaches

The total annualized costs $TAC^{(2\times max)}$ for the $(2\times max)$-reliable design $(x^f)^{(2\times max)}$ lie 5.6 % above the costs for the nominal design $\hat{x}^f$. Even though the $(2 \times max)$-reliable design $(x^f)^{(2\times max)}$ involves the highest increase in costs, reliability for all possible failures cannot be guaranteed in general without possibly high overproduction. The total annualized costs for the designs of the $(n - 1^{max})$-reliable approach and of the $(n-1)$-reliable approach are only 3.9 % higher than costs for the nominal design $\hat{x}^f$. Thus, both $(n - 1^{max})$-reliability and $(n - 1)$-reliability save 30.4 % of the additional costs for a reliable system.

Annualized operational costs $OPEX$ are nearly identical for all designs with only 0.8 % variation. The values take 5.25 Mio. €/a for the nominal problem and $(2 \times max)$-reliability, and 5.29 Mio. €/a for the proposed reliability approaches. Significant differences can be observed in the investment costs INV (Fig. 5.3): The nominal design induces 3.28 Mio. € investment costs, whereas the $(2 \times max)$-reliable design induces 4.42 Mio. € corresponding to an increase of 34.8 %. In contrast, the $(n - 1^{max})$-reliable and $(n - 1)$-reliable designs involve lower additional investment costs compared to the nominal design $\hat{x}^f$, i. e., 0.71 Mio. € corresponding to an increase of 21.6 %. Compared to the heuristic $(2 \times max)$-reliable approach, the $(n - 1^{max})$-reliable and $(n - 1)$-reliable designs save 37.7 % of the increase in investment costs. Especially in industry, low up-front costs are attractive.

The results show that cost-efficient reliable design options can be identified with both proposed reliability approaches.

5.2.2 Assessment of reliability approaches

In order to assess our approaches, we vary the demand time series by ±5 % using latin-hypercube sampling (McKay et al., 2000). Even for such slight perturbations, the obtained optimal solutions and the performance of MILP solvers can vary dramatically. Thus, we consider 10 varied instances of our original demand time series in the following.

For the 10 instances and for the original time series, we investigate the deviation of the total annualized costs from the nominal total annualized costs TAC^{nom} as well as the CPU time for computation (Tab. 5.1).

Table 5.1: Deviation of the reliable total annualized costs from the nominal total annualized costs TAC^{nom} as well as the CPU time for the original demand time series and the additional 10 instances

instance	increase of TAC^{nom} in %			CPU time in s			
	$2 \times max$	$n - 1^{max}$	$n - 1$	nom	$2 \times max$	$n - 1^{max}$	$n - 1$
original	5.6	3.9	3.9	5.7	5.8	15.3	299.3
1	5.7	3.9	3.9	6.0	6.0	15.3	506.8
2	5.3	3.8	3.8	5.6	5.7	15.2	388.1
3	5.6	3.9	3.9	5.6	5.6	15.0	555.7
4	5.6	3.9	3.9	5.6	5.6	15.1	556.4
5	5.9	4.0	4.0	5.7	5.8	15.2	369.4
6	5.7	3.9	3.9	5.7	5.8	14.9	315.7
7	5.6	3.9	3.9	5.6	5.7	15.0	492.1
8	5.5	3.7	3.7	5.7	5.7	15.1	372.6
9	5.7	3.8	3.8	5.6	5.7	15.2	407.8
10	5.6	3.9	3.9	5.7	5.8	15.1	559.3

The results show that the costs for the $(n - 1^{max})$-reliable design $(x^f)^{(n-1^{max})}$ and the $(n - 1)$-reliable design $(x^f)^{(n-1)}$ are equal for all instances. Both proposed reliability approaches reduce the total annualized costs compared to the heuristic $(2 \times max)$-reliable approach. For all instances, reliability can be achieved with a low increase of total annualized costs, i. e., 3.9 % on average. Furthermore, the shares of annualized operational costs, maintenance costs, and investment costs are similar to the original time series (Fig. 5.3). The results show that the increase in investment costs for a reliable design can be reduced by up to 45.7 % compared to the heuristic

($2 \times max$)-reliability. The annualized operational costs and the investment costs including their corresponding variation for the instances are presented in Fig. B.2. The variation of the demands shows that the expected additional costs for a reliable design are low if ($n - 1^{max}$)-reliability or ($n - 1$)-reliability is employed.

Computational times of the different approaches (Tab. 5.1) show the advantage of the ($n - 1^{max}$)-reliable approach: The ($n - 1^{max}$)-reliable approach is more than 19.6 times faster than the ($n - 1$)-reliable approach. Moreover, the ($n - 1^{max}$)-reliable approach is only maximal 2.7 times slower than the ($2 \times max$)-reliable approach which identifies designs involving high additional costs. Thus, the ($n - 1^{max}$)-reliable approach clearly outperforms ($n - 1$)-reliability in computational time and ($2 \times max$)-reliability in expected costs for a reliable design.

The differing computational times of the approaches are the result of their respective model scales (Tab. 5.2).

Table 5.2: Model scales of approaches: number of equations, continuous variables, and binary variables after presolve of the nominal problem and the three reliability approaches; computational time is again provided for comparison

approach	equations	continuous variables	binary variables	CPU time in s
nom	657	264	102	5.7
$2 \times max$	678	278	111	5.8
$n - 1^{max}$	1 435	491	227	15.3
$n - 1$	31 852	9 220	4 774	299.3

The values for model scale of ($2 \times max$)-reliability comprise values for the nominal problem plus values for additional optimization of operation with the added largest component. The ($n - 1^{max}$)-reliability approach additionally involves 2 indicator variables in GAMS for selecting whether Eq. (B.1) or Eq. (B.4) is active.

Analyzing the instances of the demand time series shows that 4 of the 10 instances lead to ($n - 1^{max}$)-reliable designs $(x^f)^{(n-1^{max})}$ which cannot cover the energy demands exactly if 1 component fails (Tab. 5.3). We call these solutions *inexact*. All inexact ($n - 1^{max}$)-reliable designs suffer from the same shortcoming: If a small cooling component fails, overproduction is necessary to guarantee sufficient energy supply in the time step with minimal cooling demand.

Table 5.3: Necessary overproduction for all instances leading to inexact $(n-1^{max})$-reliable designs; overproduction only occurs in the time step with minimal cooling demand

instance	overproduction in kW	overproduction in %
4	24	2.6
5	58.6	6.4
7	49.1	5.6
9	107.3	12.3

In order to analyze the reason for overproduction, we take a closer look at a typical inexact $(n-1^{max})$-reliable design. Fig. 5.4 shows the designs for instance 7 (Tab. B.1) of the varied demand time series.

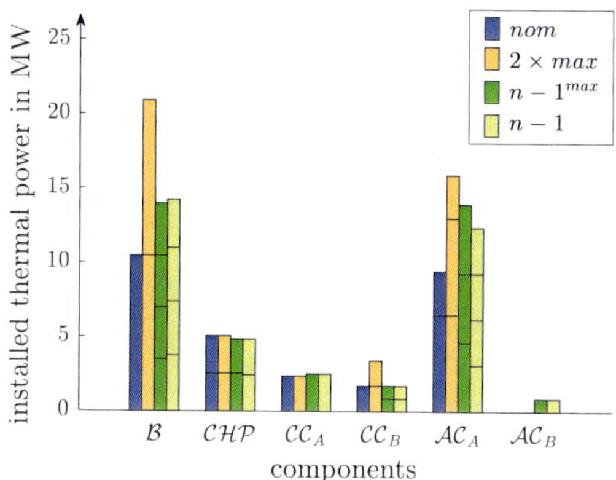

Figure 5.4: Selected components and their installed thermal power for instance 7 of the varied demand time series; components of nominal (nom) design $\hat{x}^f$, of $(2\times max)$-reliable design $(x^f)^{(2\times max)}$, of $(n-1^{max})$-reliable design $(x^f)^{(n-1^{max})}$, and of $(n-1)$-reliable design $(x^f)^{(n-1)}$ for each technology; $\mathcal{B}$ boiler, $\mathcal{CHP}$ combined heat and power engine, $\mathcal{CC}_A$ and $\mathcal{CC}_B$ compression chillers, and $\mathcal{AC}_A$ and $\mathcal{AC}_B$ absorption chillers installed on Site A and Site B, respectively

We focus on the differences between the $(n-1^{max})$-reliable design $(x^f)^{(n-1^{max})}$ and the $(n-1)$-reliable design $(x^f)^{(n-1)}$. In the $(n-1^{max})$-reliable design $(x^f)^{(n-1^{max})}$, the boilers are slightly smaller than in the $(n-1)$-reliable design $(x^f)^{(n-1)}$. However, the actual difference leading to overproduction is the larger sizing and the smaller number of absorption chillers on Site A ($\mathcal{AC}_A$) in the ($n - 1^{max}$)-reliable design $(x^f)^{(n-1^{max})}$: If the small and unique compression chiller on Site A ($\mathcal{CC}_A$) fails, the minimal part load of the replacing absorption chillers is too high to cover the minimal cooling demand exactly. However, in instance 7, only 49.1 kW of cooling are additionally produced which corresponds to 5.6 % of the needed cooling supply in this time step (Tab. 5.3). The occurring overproduction could be avoided by running 1 absorption chiller in part load with 18.9 % of the nominal installed power.

For the remaining inexact solutions, the reason for overproduction is the same. To avoid overproduction for the remaining inexact solutions, it is sufficient to decrease the part load of the absorption chiller to a minimum of 17.8% of the nominal installed power. This required part load is below the specified minimal part load of 20 % for the selected absorption chillers (Section 4.3). However, in practice, a slight decrease of the minimal part load or a slight overproduction for a short time might be justifiable in emergencies.

The analysis shows that overproduction is only a minor problem in the considered case study. Nevertheless, we want to introduce an idea how the ($n - 1^{max}$)-reliable approach can easily be further improved if desired: Just as for the failure of the largest component ensuring covering maximal demands, it is possible to add further constraints ensuring that the smallest demands can be covered exactly without the component with the smallest part load. As in the original ($n - 1^{max}$)-reliable approach, both heating and cooling circuits need to be considered in the constraints. Such an extended approach would reduce the risk of solutions leading to overproduction; however, inexact solutions involving overproduction might still occur.

The case study shows that the ($n - 1^{max}$)-reliable approach performs well even without the introduced idea for extension. The increase in computational time compared to the nominal problem is low (Tab. 5.1). Moreover, ($n - 1^{max}$)-reliability enables the design to ensure sufficient energy supply during failure with only slightly increased total annualized costs.

5.3 Conclusions

In the design of DESS, reliability of the system is an important but often neglected issue. We propose an approach to identify exact ($n - 1$)-reliable designs for DESS.

Our $(n-1)$-reliable approach ensures reliable energy supply during the failure of 1 component at any time. Thus, the approach also enables to maintain components at any time if no failure occurs simultaneously.

The number of additional constraints in the $(n-1)$-reliable approach depends on the product of the number of potentially installed components and the number of time steps leading to a polynomial increase of the problem complexity. Thus, the exact approach might lead to high computational effort. For this reason, we propose also a computationally-efficient inexact approach, called $(n - 1^{max})$-reliability which ensures that all demands can still be covered if 1 arbitrary component fails but allows for overproduction. The computationally-efficient $(n - 1^{max})$-reliable approach involves only few additional constraints and the increase in computational time is low. Therefore, $(n - 1^{max})$-reliability would allow increasing model accuracy or coupling reliability with other optimization approaches with increased computational time, as performed in this thesis in Chapter 8.

An industrial real-world case study shows that pragmatic heuristics from industry applications, such as installing redundant components by rules of thumb, lead to unnecessary high additional investment costs for reliability. Employing both the proposed exact and inexact reliability approaches, we obtain well-performing solutions during failure and achieve savings of 30.4 % in the additional total annualized costs compared to pragmatic heuristics from industry applications. The increase in the investment costs for a reliable design can even be reduced by 37.7 % employing the proposed reliability approaches. Reliable designs of the proposed $(n-1)$-reliable approach and the $(n - 1^{max})$-reliable approach lead to an increase in total annual costs of only 3.9 % compared to the total annualized costs of an optimal system design without providing any reliability. Thus, our proposed approaches identify optimal reliable designs for DESS at low additional costs. The exact $(n-1)$-reliable approach can ensure reliability while $(n - 1^{max})$-reliability provides a well-performing inexact approach which is easy to implement and which involves low additional computational effort.

Employing the proposed reliability approaches ensures that any component can be maintained at any time and ensures sufficient energy supply during failure of any component. Thereby, unpredictable costs can be avoided and security in industrial applications can be improved significantly. For the integration into the be-rebust framework, the $(n - 1^{max})$-reliability approach is highly suitable due to the short computational time for solving the reliability problem with high solution quality.

Chapter 6

Robustness against uncertain input parameters

Uncertainty of input parameters during optimization is a common problem in the synthesis of DESS. In this chapter, we introduce two approaches for regarding uncertainties of input parameters into the synthesis of DESS. First, we introduce the ***Two-stage Robustness Trade-off (TRusT)*** approach. The TRusT approach is an appropriate tool to design cost-efficient and secure DESS. Second, to identify sustainable designs while regarding uncertainty of input parameters, we introduce the concept of *minmax **Robust Multi-objective Optimization (RoMO)*** (Ehrgott et al., 2014) into DESS optimization. The proposed approaches allow for incorporating uncertainty of input parameters into the be-rebust framework (Fig. 6.1).

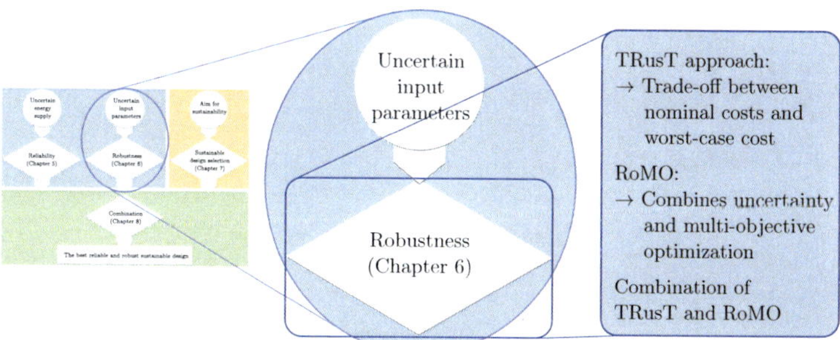

Figure 6.1: Overview Chapter 6 – TRusT approach and RoMO for robustness against uncertain input parameters

The contribution of the TRusT approach to this thesis is to show that a robust design which *ensures* sufficient energy supply can be cost-efficient at the same time. The advantages of the TRusT approach are twofold: First, the approach highlights the trade-off between nominal costs and worst-case costs for the robust design (Fig. 6.2). As a reminder, the nominal costs represent costs obtained when uncertainties are neglected and perfect foresight is assumed. Second, the TRusT approach guarantees sufficient energy supply as the approach takes into account uncertain input parameters. Existing approaches often make compromises: The robustness of the system is reduced in order to receive a less expensive design, as discussed in Section 2.4.2.

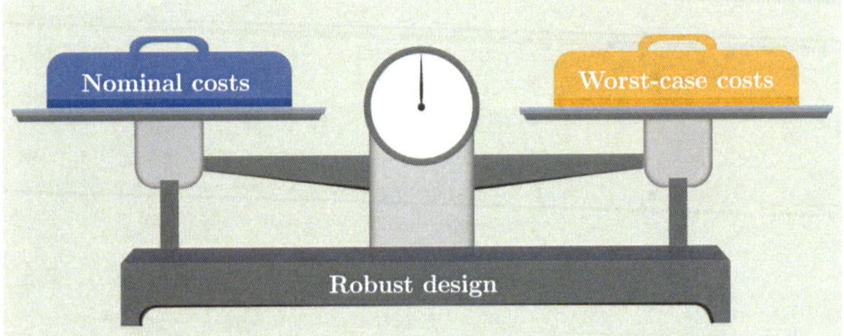

Figure 6.2: TRusT approach regards uncertainty of input parameters and reveals the trade-off between nominal costs and worst-case costs

In the optimal design of DESS while taking into account the aspects of sustainability, we incorporate uncertainty of input parameters by applying the concept of RoMO (Fig. 6.3). In particular, we transfer the general mathematical concept proposed by Ehrgott et al. (2014) to the considered DESS optimization model introduced in Section 4.1 which includes typical characteristics of DESS. For the synthesis of sustainable DESS which involves multiple criteria for decision making, RoMO allows for regarding uncertain input parameters.

Both, the TRusT approach and RoMO are applicable to any other two-stage synthesis problem and are not restricted to DESS optimization.

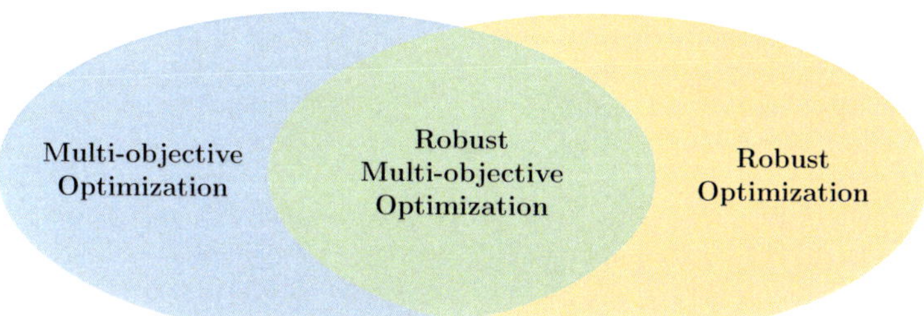

Figure 6.3: RoMO identifies sustainable designs while taking uncertainty of input parameters into account by employing multi-objective and robust optimization

This chapter is structured as follows: In Section 6.1.1, we explain the idea of the TRusT approach before introducing the formulation for minmax robust optimization in Section 6.1.2 which is employed in the TRusT approach (Section 6.1.3) and in RoMO (Section 6.2.1). In Section 6.1.4, we state the problem formulation of the TRusT approach for DESS. The TRusT approach is applied to the real-world problem and the sensitivity on uncertain input parameters is analyzed in Section 6.1.5. A brief introduction to RoMO theory is given in Section 6.2.1. The theoretical introduction is followed by a discussion on simplifications resulting from the kind of uncertainties which are common in energy system optimization. In Section 6.2.2, we propose the resulting minmax robust problem formulation of the sustainable DESS design problem and adapt this formulation to specific characteristics of DESS to allow for easy solution computation. The proposed formulation is applied to the real-world case study in Section 6.2.3. To yield robust sustainable designs with minimal nominal costs, we combine the TRusT approach and RoMO in Section 6.3. A brief summary and conclusions of Chapter 6 are given in Section 6.4.

6.1 Two-stage Robustness Trade-off (TRusT) of distributed energy supply systems

Our approach builds upon the classical concept of strictly (also called minmax) robust optimization stated by Soyster (1973). Robust optimization can be efficiently applied to DESS design using MILP optimization. We extend the concept of minmax robustness and introduce the TRusT approach. The TRusT approach exploits the two-stage nature of DESS design: A two-stage bi-objective optimization is employed studying the nominal cost and the worst-case costs simultaneously. Both objective functions employ *identical* first-stage variables representing design decisions. The second-stage variables, however, are adapted *separately* for the nominal and the robust objective function (Section 6.1.1). Thus, the design decisions ensure that the system is feasible for every scenario. Operation can thus be adapted to each specific case, e. g., the nominal or the worst-case scenario.

A related *bi-criteria approach to robust optimization* has been introduced by Chassein and Goerigk (2016) for *single*-stage problems with uncertainties in the objective function. In their approach, the nominal and the robust objective function are minimized. Schöbel (2014) generalizes the concept of *light robustness* (Fischetti and Monaci, 2009) which selects solutions with the highest robustness among all solutions within a certain range of the nominal optimal objective function value. They show that the trade-off between robustness and nominal quality yields promising solutions. The trade-off between minmax robustness and minmax regret based on adaptive robust optimization is investigated by Ning and You (2018) for planning and scheduling problems. For stochastic programming, a similar concept has recently been proposed Fuentes-Cortés et al. (2016): A stochastic analysis is performed for the expected and worst-case costs of residential cogeneration systems.

The proposed TRusT problem does not depend on unknown probability distributions and is formulated as a bi-objective MILP. Thus, the problem is solvable in reasonable time without implementing special solution algorithms which allows easy application to DESS design problems. Thereby, the TRusT approach increases the acceptance for practical applications. Furthermore, we show that the identified robust optimal designs of the DESS guarantee sufficient energy supply and, nevertheless, can be implemented at low additional costs. Thus, the results have significant practical relevance.

Major parts of this section are reproduced by permission of Elsevier from:

> Majewski, D. E., Lampe, M., Voll, P., Bardow, A. (2017a) TRusT: A Two-stage Robustness Trade-off approach for the design of decentralized energy supply systems. *Energy*, 118:590–599

Contribution report: Principal author, development of approach, implementation of the approach, analysis of results, writing the draft

6.1.1 Idea of the TRusT approach

In this section, the idea of the TRusT approach is introduced for the design of DESS. We incorporate uncertainties by taking advantage of the two-stage nature (Section 2.2): The first-stage variables x^f correspond to the design variables which determine structure and sizing. The second-stage variables x^s fix the operation and can be adapted to the occurring scenario. Allowing for uncertainties in the energy demands has an impact not only on the operation of the DESS but also on the design.

The *robust* objective function $f^{rob/rob}$ minimizes the worst-case costs while the constraints must be fulfilled for every possible scenario: This leads to a robust design x^f and robust operation x^s and, thus, to the classical problem of robust optimization (Soyster, 1973) for DESS design. The robust optimization is represented by the right branch in Fig. 6.4. The resulting robust optimal solution is generally conservative.

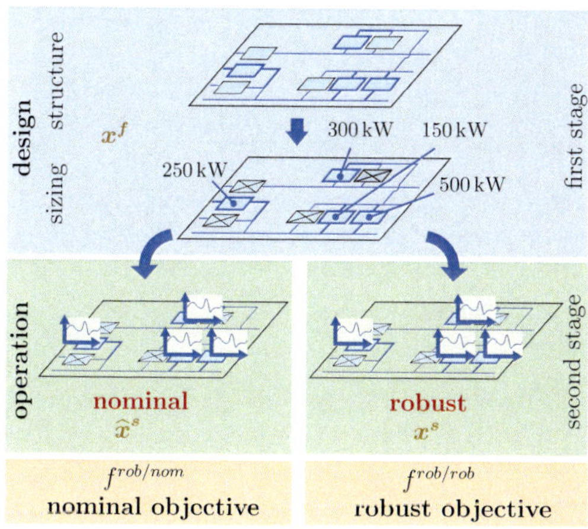

Figure 6.4: Illustration of the TRusT approach based on bi-objective optimization with objectives $f^{rob/nom}$ and $f^{rob/rob}$. The nominal objective $f^{rob/nom}$ considers the robust design determined by robust first-stage variables x^f and nominal operation represented by the second-stage variables $\hat{x}^s$; the robust objective $f^{rob/rob}$ considers robustness in both stages, i. e., for design variables x^f and for operational variables x^s.

Conservative solutions are often desired for energy supply systems, as the sufficient energy supply is mandatory (Kaundinya et al., 2009). However, there is a trade-off between the degree of conservatism of the design and the resulting costs. Thus, the TRusT approach complements robust optimization by taking into account also the *nominal scenario* (left branch in Fig. 6.4). The nominal scenario corresponds to the best available prediction for the uncertain parameters which would have been employed in the nominal problem with assumed perfect foresight.

The bi-objective TRusT approach integrates the nominal evaluation into the minmax robust concept by optimizing the nominal objective function $f^{rob/nom}$ *and* the robust objective function $f^{rob/rob}$. However, not only the robust objective function $f^{rob/rob}$ but also the nominal objective function $f^{rob/nom}$ employs a robust design x^f to ensure sufficient energy supply. Hence, using the nominal objective $f^{rob/nom}$, a robust design x^f must be found which is also cost efficient for the nominal scenario. The two objective functions evaluate the performance of the *same* robust design x^f in the nominal and in the worst-case scenario. Thus, the TRusT approach identifies the trade-off between conservativeness and everyday costs.

6.1.2 Minmax robust optimization

The TRusT approach is based on minmax robust optimization (Soyster, 1973). Hence, this section gives a brief introduction to the concept of minmax robust optimization based on Ben-Tal and Nemirovski (1999).

Assuming that potentially uncertain parameters ξ are known with certainty, the optimization problem $\left(\mathcal{OP}_\xi\right)$ can be defined by

$$\left(\mathcal{OP}_\xi\right) \quad \begin{array}{ll} \min & f(x,\xi) \\ \text{s.t.} & x \in \mathcal{X}(\xi), \end{array}$$

where f specifies the objective function and the $\mathcal{X}(\xi)$ the set of feasible solutions.

In real life, the parameters ξ in the optimization are often uncertain. Each occurring set of uncertain parameters is called a *scenario*. The union of all scenarios is represented by the *uncertainty set* $\mathcal{U}$. Taking all uncertain parameters into account leads to an *uncertain optimization problem* (Ben-Tal and Nemirovski, 1999) defined by the family

$$(\mathcal{OP}_\mathcal{U}) := \left(\left(\mathcal{OP}_\xi\right), \xi \in \mathcal{U}\right).$$

The problem $\left(\mathcal{OP}_\xi\right)$ is the deterministic optimization problem resulting for one particular scenario $\xi \in \mathcal{U}$, i. e., for each uncertain parameter, a certain value is assumed. The uncertain problem $(\mathcal{OP}_\mathcal{U})$ represents the union of all sub-problems $\left(\mathcal{OP}_\xi\right)$ over all scenarios; thus, all scenarios of the uncertainty set $\mathcal{U}$ are considered.

The *robust counterpart* $(\mathcal{OP}_\mathcal{RC})$ of the uncertain problem $(\mathcal{OP}_\mathcal{U})$ minimizes the maximal possible objective function value while the constraints of each Problem $\left(\mathcal{OP}_\xi\right)$ are forced to hold for all scenarios (Ben-Tal and Nemirovski, 1999):

$$
\begin{array}{llll}
(\mathcal{OP}_\mathcal{RC}) & \min & \sup_{\xi \in \mathcal{U}} f(x,\xi) & \\
 & \text{s.t.} & x \in \mathcal{X}(\xi) & \forall \xi \in \mathcal{U}.
\end{array}
$$

For purpose of solvability, we consider the equivalent formulation:

$$
\begin{array}{llll}
(\mathcal{OP}^\alpha_\mathcal{RC}) & \min & \alpha & \\
 & \text{s.t.} & f(x,\xi) \leq \alpha & \forall \xi \in \mathcal{U} \\
 & & x \in \mathcal{X}(\xi) & \forall \xi \in \mathcal{U} \\
 & & \alpha \in \mathbb{R}. &
\end{array}
$$

The equivalence holds because $(\mathcal{OP}^\alpha_\mathcal{RC})$ minimizes the auxiliary variable α, which is an upper bound for the original objective function $f(x,\xi)$ no matter which scenario occurs (Ben-Tal and Nemirovski, 1999). An optimal solution of $(\mathcal{OP}_\mathcal{RC})$ is *robust optimal* for the uncertain problem $(\mathcal{OP}_\mathcal{U})$.

In the robust counterpart $(\mathcal{OP}_\mathcal{RC})$, the solution must be feasible for every scenario ξ. The objective determines the maximal cost over all scenarios. Thus, the resulting optimal solution is feasible for all scenarios and minimizes the worst-case costs.

6.1.3 The TRusT approach

The TRusT approach identifies robust design options ensuring sufficient energy supply. Cost-efficient design options can be chosen by analyzing the trade-off between nominal costs and worst-case, i. e., robust, costs. Thus, the proposed TRusT approach allows the designer to find solutions which are both, robust *and* cost-efficient.

The TRusT approach applies bi-objective optimization to a two-stage uncertain problem. The bi-criteria optimization problem of the TRusT approach employs identical first-stage variables x^f for both objective functions (Section 6.1.1). The second stage is determined by the variables $\widehat{x}^s$ and x^s for the nominal and the robust objective function, respectively. The corresponding *TRusT problem* $\left(\mathcal{OP}_{TRusT}\right)$ is defined by

$$
\begin{aligned}
\left(\mathcal{OP}_{TRusT}\right) \quad & \min_{x^f,\widehat{x}^s,x^s} \begin{pmatrix} f\left(x^f,\widehat{x}^s,\widehat{\xi}\right) \\ \sup_{\xi\in\mathcal{U}} f\left(x^f,x^s,\xi\right) \end{pmatrix} & \\
& \text{s. t.} \quad x^f \in \mathcal{X}^f(\xi) & \forall \xi \in \mathcal{U} \\
& \qquad \widehat{x}^s \in \mathcal{X}^s\left(x^f,\widehat{\xi}\right) & \\
& \qquad x^s \in \mathcal{X}^s(x^f,\xi) & \forall \xi \in \mathcal{U}\,,
\end{aligned}
$$

with uncertainty set $\mathcal{U}$, and functions $f : \mathcal{X}^f \times \mathcal{X}^s \times \mathcal{U} \to \mathbb{R}$ and feasible set of solutions for the first stage $\mathcal{X}^f$ and the second stage $\mathcal{X}^s$. The objective function $f\left(x^f,\widehat{x}^s,\widehat{\xi}\right)$ corresponds to the nominal objective $f^{rob/nom}$ and $\sup_{\xi\in\mathcal{U}} f\left(x^f,x^s,\xi\right)$ to the robust objective $f^{rob/rob}$, respectively (Fig. 6.4). The Pareto front of this problem is called *TRusT curve* and contains all efficient solutions of problem $\left(\mathcal{OP}_{TRusT}\right)$, i. e., solutions which cannot be improved for both objective functions simultaneously. The supremum in problem $\left(\mathcal{OP}_{TRusT}\right)$ can be substituted by the auxiliary variable α in the same way as in problem $(\mathcal{OP}^{\alpha}_{\mathcal{RC}})$.

The proposed TRusT approach provides the trade-off between the nominal objective function $f\left(x^f,\widehat{x}^s,\widehat{\xi}\right)$ and the robust objective function $\sup_{\xi\in\mathcal{U}} f\left(x^f,x^s,\xi\right)$. Both objective functions employ the same robust first-stage variables x^f for the structure, whereas the second-stage variables $\widehat{x}^s$ and x^s are adjusted independently. The TRusT approach is not confined to the design of DESS but can be applied to any two-stage optimization problem.

6.1.4 TRusT formulation for distributed energy supply systems

When stating the TRusT problem for design of DESS, we are potentially confronted with two problems: First, the problem is intrinsically infeasible and, second, there is an infinite number of constraints. Thus, the problem would not be solvable. In the following, we show why these two problems occur and how we can adapt the TRusT problem.

According to minmax robust optimization, we enforce feasibility of a solution for *every* scenario in the TRusT problem: There is no setting of the operation variables x^s such that all possible energy demands can be fulfilled exactly *for all scenarios*. In practice, the operation is fixed once the occurring demands are known. However, we need to fix the operation already during the optimization of the design, since the operation influences the optimization of the design (Frangopoulos et al., 2002). To be able to adapt the operation to any demands which might occur, it is necessary to introduce operation variables for each scenario which might occur. For a discrete uncertainty set, this can easily be implemented. Since we consider interval-based uncertainty in this thesis, adapting the operation to every scenario would involve an infinite number of variables. This results from the infinite number of demand scenarios which might occur within the given intervals. Thus, it is not viable to allow the exact adaptation of the operation to every scenario. For this reason, we need to adapt the energy balances.

To formulate a feasible optimization problem, we first replace equalities in the energy balances by inequalities:

$$\begin{aligned} \sum_{k \in \mathcal{B} \cup \mathcal{CHP}} \dot{V}_{kt} - \sum_{k \in \mathcal{AC}} \dot{U}_{kt} &\geq \tilde{\dot{E}}_t^h \quad \forall t \in \mathcal{T}, \xi \in \mathcal{U} \\ \sum_{k \in \mathcal{AC} \cup \mathcal{CC}} \dot{V}_{kt} &\geq \tilde{\dot{E}}_t^c \quad \forall t \in \mathcal{T}, \xi \in \mathcal{U} \\ \sum_{k \in \mathcal{CHP}} \dot{V}_{kt}^{el} - \sum_{k \in \mathcal{CC}} \dot{U}_{kt}^{el} + \dot{U}_t^{el,buy} - \dot{V}_t^{el,sell} &\geq \tilde{\dot{E}}_t^{el} \quad \forall t \in \mathcal{T}, \xi \in \mathcal{U}. \end{aligned} \tag{6.1}$$

Boilers and combined heat and power engines provide heating energy $\sum_{k \in \mathcal{B} \cup \mathcal{CHP}} \dot{V}_{kt}$ to cover the uncertain demand $\tilde{\dot{E}}_t^h$. Furthermore, boilers and combined heat and power engines supply the absorption chillers $\sum_{k \in \mathcal{AC}} \dot{U}_{kt}$. The output of the absorption chillers and compression chillers $\sum_{k \in \mathcal{AC} \cup \mathcal{CC}} \dot{V}_{kt}$ is used to cover the uncertain cooling demand $\tilde{\dot{E}}_t^c$. To cover the uncertain electricity demand $\tilde{\dot{E}}_t^{el}$ and to supply the compression chillers $\sum_{k \in \mathcal{CC}} \dot{U}_{kt}^{el}$, electricity flow provided by the combined heat and power engines $\sum_{k \in \mathcal{CHP}} \dot{V}_{kt}^{el}$ and the electricity grid $\dot{U}_t^{el,buy}$ is used. Electricity flow can also be fed into the grid $\dot{V}_t^{el,sell}$. Relaxing the energy balances to inequalities ensures that at least the required energy is provided.

However, the problem is still not solvable, since it comprises an infinite number of constraints as the uncertainty set $\mathcal{U}$ has an infinite number of elements. However, it is sufficient to consider only the upper bound of the uncertain demands. For all types of demands, $\hat{\dot{E}}_t + \delta_t^{\dot{E}} \geq \tilde{\dot{E}}_t$ holds for all scenarios $\xi \in \mathcal{U}$. As a result, all demands below the upper bound $\hat{\dot{E}}_t + \delta_t^{\dot{E}}$ can be covered when the upper bound can be covered. Thus, only the upper bound is decisive and any other constraints can be eliminated as redundant:

$$
\begin{aligned}
\sum_{k\in\mathcal{B}\cup\mathcal{CHP}} \dot{V}_{kt} - \sum_{k\in\mathcal{AC}} \dot{U}_{kt} &\geq \hat{\dot{E}}_t^h + \delta_t^{\dot{E}h} \quad \forall t \in \mathcal{T} \\
\sum_{k\in\mathcal{AC}\cup\mathcal{CC}} \dot{V}_{kt} &\geq \hat{\dot{E}}_t^c + \delta_t^{\dot{E}c} \quad \forall t \in \mathcal{T} \\
\sum_{k\in\mathcal{CHP}} \dot{V}_{kt}^{el} - \sum_{k\in\mathcal{CC}} \dot{U}_{kt}^{el} + \dot{U}_t^{el,buy} - \dot{V}_t^{el,sell} &\geq \hat{\dot{E}}_t^{el} + \delta_t^{\dot{E}e} \quad \forall t \in \mathcal{T}.
\end{aligned}
\tag{6.2}
$$

There might exist scenarios for which the *total* heating demand $\tilde{\dot{E}}_t^{h,tot}$ (including heating energy for operating the absorption chillers $\sum_{k\in\mathcal{AC}} \dot{U}_{kt}$) is higher than the total heating demand for the upper bound $\hat{\dot{E}}_t^h + \delta_t^{\dot{E}h} + \sum_{k\in\mathcal{AC}} \dot{U}_{kt}$. However, in this case, the system could still be operated in the same operation mode as employed for covering the upper bound of the heating demands $\hat{\dot{E}}_t^h + \delta_t^{\dot{E}h}$. In this operational mode, the absorption chillers $\sum_{k\in\mathcal{AC}} \dot{U}_{kt}$ are replaced by compression chillers. Hence, sufficient energy supply can be ensured even though overproduction might be necessary.

We reduce overproduction by making use of the two-stage nature of the problem: Since it is not possible to ensure exactly fulfilled energy balances for all considered scenarios (as described in the beginning of this section), we only enforce that the operation can be adapted such that the energy balances hold exactly for the nominal demands and for the lower bounds of the demands. As a result, the energy system is designed to cover small demands exactly leading to a design with reduced overproduction. To include the nominal demands and the lower bounds of the demands, we introduce two additional constraints for each energy balance and new operation variables. As an example, the additional constraints for heating supply are

$$
\begin{aligned}
\sum_{k\in\mathcal{B}\cup\mathcal{CHP}} \hat{\dot{V}}_{kt} - \sum_{k\in\mathcal{AC}} \hat{\dot{U}}_{kt} &= \hat{\dot{E}}_t^h \quad \forall t \in \mathcal{T} \\
\sum_{k\in\mathcal{B}\cup\mathcal{CHP}} \underline{\dot{V}}_{kt} - \sum_{k\in\mathcal{AC}} \underline{\dot{U}}_{kt} &= \hat{\dot{E}}_t^h - \delta_t^{\dot{E}h} \quad \forall t \in \mathcal{T},
\end{aligned}
\tag{6.3}
$$

where $\hat{\dot{V}}_{kt}$ and $\hat{\dot{U}}_{kt}$ as well as $\underline{\dot{V}}_{kt}$ and $\underline{\dot{U}}_{kt}$ are additional operation variables to adapt the operation to the nominal demands and to the lower bounds of the demands. However, the new variables for the lower bounds $\underline{\dot{V}}_{kt}$ and $\underline{\dot{U}}_{kt}$ do not have a direct influence on the objective functions.

The problem still contains an infinite number of constraints induced by the robust objective function: In order to eliminate the supremum, the auxiliary variable α

is introduced (Section 6.1.3). The auxiliary variable α limits the total annualized costs for every scenario $\xi \in \mathcal{U}$:

$$\sum_{t\in\mathcal{T}}\left[\Delta\tau_t\left(\tilde{p}^{gas}\cdot\sum_{k\in\mathcal{B}\cup\mathcal{CHP}}\dot{U}_{kt}+\tilde{p}^{el,buy}\cdot\dot{U}_t^{el,buy}-\tilde{p}^{el,sell}\cdot\dot{V}_t^{el,sell}\right)\right] + \sum_{k\in\mathcal{K}}\left(\frac{1}{PVF}+p_k^m\right)\cdot I_k\leq\alpha\quad\forall\xi\in\mathcal{U}.$$

Recall that one scenario ξ represents one combination of parameters for uncertain prices ($\tilde{p}^{gas}$, $\tilde{p}^{el,sell}$, and $\tilde{p}^{el,buy}$), for the uncertain specific global warming impact $\widetilde{GWI}^{el}$, and for uncertain energy demands $\left(\left(\tilde{\dot{E}}_t^h\right)_{t\in\mathcal{T}},\left(\tilde{\dot{E}}_t^c\right)_{t\in\mathcal{T}},\text{ and }\left(\tilde{\dot{E}}_t^{el}\right)_{t\in\mathcal{T}}\right)$ (Section 4.2).

Finally, we eliminate the resulting infinite number of constraints by inserting the upper bound for the gas price (Eq. (6.4)). This is exact because the highest price corresponds to the highest cost. The uncertain electricity costs for purchasing and selling energy vary uniformly because the price levels are correlated. As revenue from electricity sales reduce total annualized costs, not only the upper bound but also the lower bound of the electricity prices has to be taken into account. Hence, the following constraints have to be considered:

$$\sum_{t\in\mathcal{T}}\left[\Delta\tau_t\left(\hat{p}^{gas}(1+\delta^{pg})\cdot\sum_{k\in\mathcal{B}\cup\mathcal{CHP}}\dot{U}_{kt} + \hat{p}^{el,buy}(1+pe)\cdot\dot{U}_t^{el,buy}-\hat{p}^{el,sell}(1+pe)\cdot\dot{V}_t^{el,sell}\right)\right] + \sum_{k\in\mathcal{K}}\left(\frac{1}{PVF}+p_k^m\right)\cdot I_k\leq\alpha\quad\forall\, pe\in\left\{-\delta^{pe},\delta^{pe}\right\}. \tag{6.4}$$

This formulation involves two constraints: one for high and one for low electricity prices with $pe = -\delta^{pe}$ and $pe = \delta^{pe}$, respectively.

By eliminating the supremum and reducing the constraints to a finite number, all constraints and the robust objective function are reformulated to mixed-integer linear relations. The nominal objective function is already a mixed-integer linear function, so no reformulation is necessary. The resulting problem is a bi-objective MILP and can be solved with common concepts for multi-objective optimization (Ehrgott, 2005). For reference, the full problem is stated in Appendix C.1.

6.1.5 Results of the case study

In this section, the TRusT approach is applied to the nominal problem introduced in Section 4.1. Uncertainties are regarded in demands and prices as discussed in Section 4.2. In this section, the already existing components on site are taken into account. In this section, the time horizon is assumed to be $\vartheta = 10$ years. The problem is formulated in GAMS 24.3.3 (McCarl, 2014) and is solved to machine accuracy with CPLEX 12.6.0.1 (IBM Corporation, 2015) on a computer with 3.24 GHz and 64 GB RAM.

Optimal solutions for single-objective optimization

If perfect foresight is assumed, i. e., only the nominal scenario is considered, optimal total annualized costs are $(TAC^{nom})^* = 5.92$ Mio. €/a. The corresponding optimal solution requires the installation of 2 combined heat and power engines, 1 absorption chiller, and 1 compression chiller on Site A (Fig. 6.5). From the existing energy system, 1 boiler and 1 compression chiller are retained. On Site B, 1 absorption chiller is installed. This structure is referred to as *reference structure* in the following. However, if energy prices rise, the nominal optimal solution will become suboptimal. Even worse, the solution will be infeasible if the demands increase according to the uncertainties (Fig. 4.1). As a result, the energy demands could not be covered.

Sufficient energy supply is ensured for all ranges of uncertainty by the single-objective robust problem. The resulting robust optimal solution employs the same, but much larger components as the nominal optimal solution (Fig. 6.5). Thus, the robust optimal total annualized costs are $(TAC^{rob})^* = 10.63$ Mio. €/a which corresponds to an increase of 80 % compared to the nominal optimal value from the reference structure. However, the robust optimal solution considers the worst-case scenario for the demands. Therefore, the robust solution supplies a significantly lager amount of energy than the solution of the nominal problem. Even more importantly, the minmax robust problem assumes higher prices for the energy purchased. Thus, there is no common basis for a direct comparison of nominal and robust optimal costs, $(TAC^{nom})^*$ and $(TAC^{rob})^*$, respectively. To provide a sound comparison and to find an adequate trade-off between nominal quality and robustness, we apply the bi-objective TRusT approach.

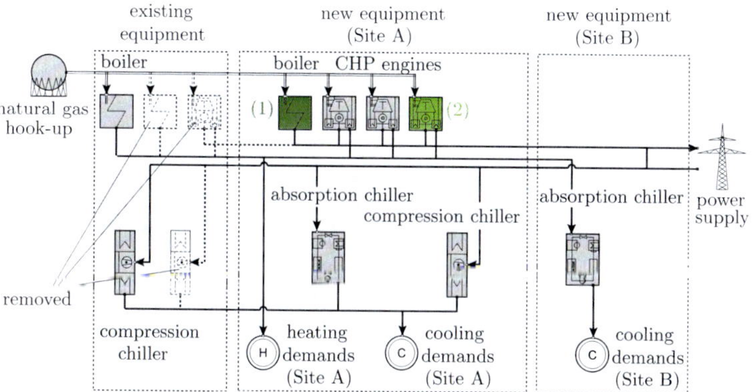

Figure 6.5: Optimal design for both, single-objective nominal and single-objective robust solution depicted in light gray (referred to as reference structure); additional dark green and green components (marked by "(1)" and "(2)", respectively) are referred to in section 6.1.5 in Fig. 6.6: "(1)" only in solutions with "×"-marks, "(2)" only in solutions marked by "∗"

Two-stage Robustness Trade-Off curve

The TRusT approach provides insight on the true costs related to robust energy supply systems. The TRusT problem for the real-world example described in Section 4.3 is solved with the augmented ε-constraint method (Mavrotas, 2009). In contrast to the single-objective nominal problem (Section 4.3), *both* objective functions are based on robust design variables. To emphasize this difference, we use the expression *robust feasible* costs if the reference might not be clear. The *TRusT curve* of the total annualized costs is the Pareto front of the robust feasible nominal total annualized costs $TAC^{rob/nom}$ and the robust feasible robust total annualized costs $TAC^{rob/rob}$. The TRusT curve of the real-world DESS design problem is shown in Fig. 6.6.

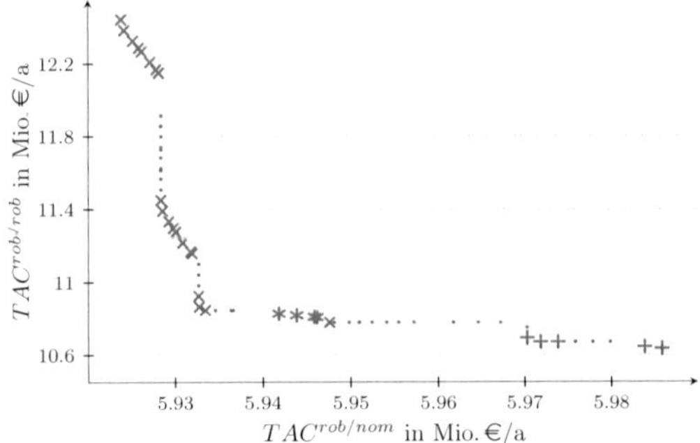

Figure 6.6: TRusT curve of the real-world problem; values corresponding to the reference structure shown in Fig. 6.5 (light gray components) are marked by "×", solutions with 1 more boiler than contained in the reference structure are marked with "∗" (and shown in dark green and marked with "(1)" in Fig. 6.5), solutions marked by "+" contain 1 additional combined heat and power engine (shown in green and marked with "(2)" in Fig. 6.5)

The anchor point on the top left-hand side (Fig. 6.6) represents the minimal robust feasible nominal total annualized costs $(TAC^{rob/nom})^*$: The solution is based on robust design variables, i. e., the design is feasible for all demands, while determining the operation for the nominal scenario. The second anchor point on the right-hand side corresponds to the minimal robust feasible robust total annualized costs $\left(TAC^{rob/rob}\right)^*$. This solution considers not only a robust design but also robust operation. Additionally, each design can exactly fulfill both the nominal demands as well as the lower bounds of the demands (Eq. (6.3)). The TRusT curve has kinks at 5.928 Mio. €/a and 5.933 Mio. €/a, respectively, where a remarkable decrease of the robust feasible robust total annual costs $TAC^{rob/rob}$ can be observed. The costs decrease because the sizing of the absorption chiller on Site A increases along with a decrease of the sizing of the compression chiller. In combination with excess heat of a combined heat and power engine, an absorption chiller is more cost efficient than a compression chiller.

Fig. 6.7 shows the robust design x^f of the anchor points and of the solution at the kink with costs of $TAC^{rob/nom} = 5.933$ Mio. €/a which provides a good trade-off.

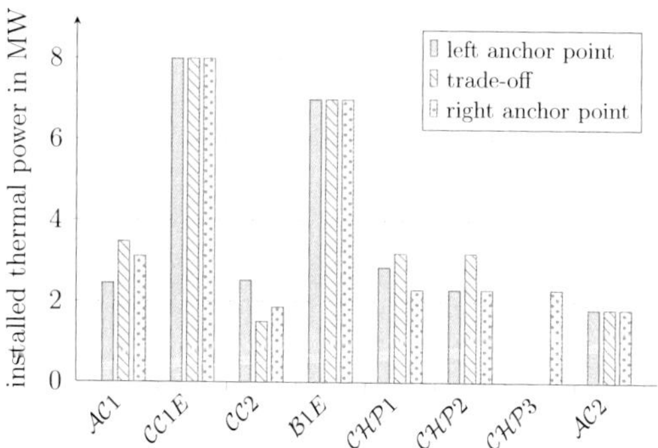

Figure 6.7: Robust design x^f of 3 efficient solutions in terms of installed thermal power for robust feasible nominal optimal total annualized costs $(TAC^{rob/nom})^*$ (left anchor point; Fig. 6.6), for the solution at the kink (see text for details) with costs of $TAC^{rob/nom} = 5.933$ Mio. €/a (trade-off), and for robust feasible robust optimal total annualized costs $(TAC^{rob/rob})^*$ (right anchor point; Fig. 6.6): $\mathcal{AC}1$, $\mathcal{AC}2$ absorption chillers, $\mathcal{CC}1E$, $\mathcal{CC}2$ compression chillers, $\mathcal{B}1E$ boiler, $\mathcal{CHP}1$, $\mathcal{CHP}2$, $\mathcal{CHP}3$ combined heat and power engines; components ending with "E" are already existing components with a fixed size; only $\mathcal{AC}2$ is placed on Site B.

There are only small variations in the sizing of the components. The main difference is that the robust optimal design x^f with minimal robust costs $(TAC^{rob/rob})^*$ selects 1 additional combined heat and power engine and, thus, provides more heat. Consequently, absorption chiller $\mathcal{AC}1$ which runs on heat is larger and compression chiller $\mathcal{CC}2$ is smaller compared to the robust optimal design for minimal nominal costs $(TAC^{rob/nom})^*$.

In Fig. 6.6, all values of the robust feasible nominal total annualized costs $TAC^{rob/nom}$ lie within a range of only 1 %. Thus, even installing the robust optimal design implies only low costs $TAC^{rob/nom}$ for the nominal conditions. The maximal deviation to the single-objective $(TAC^{nom})^*$ is only 1.2 %. The nominal solution is infeasible for the uncertain problem and represents a lower bound for all solutions including solutions from other design approaches that include uncertainties. Approaches that compromise on the security of energy supply might lead to a more cost-efficient design than the TRusT solutions. However, the upper bound of the potential savings is only 1.2 % of the robust costs which corresponds to 70 000 €/a. A reason for the small increase of the costs for a robust design is that peak demands are already considered in the nominal problem. The robust design does not cause a further large increase of the costs. Thus, using peak demands in conventional optimization results in similar additional costs as a robust design. However, the conventional design does *not* ensure sufficient energy supply. The comparison of the nominal optimal total annualized costs $(TAC^{nom})^*$ with the costs $(TAC^{rob/nom})^*$ of a robust system for the nominal conditions is more adequate than examining the deviation between nominal optimal costs $(TAC^{nom})^*$ and the robust optimal costs $(TAC^{rob})^*$ of the single-objective optimization problems which is 80 %. The comparison of the nominal optimal costs $(TAC^{nom})^*$ with the robust optimal costs $(TAC^{rob})^*$ does not provide the sound basis for a comparison of the designed energy system because the single-objective robust problem assumes higher demands and higher prices. The higher energy prices would also increase the costs of the nominal optimal design. Furthermore, the nominal optimal design is not feasible for all scenarios, thus, a lack of energy supply may arise for several scenarios if the nominal optimal solution was implemented.

For the robust designs from the TRusT curve, the range is quite small for the nominal costs, whereas the robust costs vary significantly: The range of the robust total annualized costs $TAC^{rob/rob}$ is 17 % (Fig. 6.6). This implies that a robust design that is cost efficient for the nominal scenario can cause significantly higher costs if an unexpected scenario occurs.

The case study shows that the TRusT approach provides robust designs at low additional costs in the nominal scenario. The constraints in Eq. (6.2) and in Eq. (6.3) ensure that every robust design can cover at least any demand in the uncertainty interval and it is able to cover exactly the nominal demands and the lower bound of each demand. The solutions at kinks of the TRusT curve provide a good trade-off: Slightly higher costs for the nominal scenario lead to significantly lower costs in the robust case. It is remarkable that the full range of robust efficient design is not expensive, when operated for the nominal scenario. Even the maximal deviation to the nominal optimal total annualized costs $(TAC^{nom})^*$ is only 1.2 %.

Sensitivity analysis of the TRusT curve

In the previous section, we consider the uncertainty interval to be explicitly known. However, in real life, lower and upper bounds are in general unknown themselves

(Zhao and You, 2019). Thus, we analyze the sensitivity of the total annualized costs with respect to the bounds of the demand uncertainties. In the following, we vary the lower and upper bounds of the uncertainty intervals for the demands by scaling the uncertainty size $\delta_t^{\dot{E}h}$, $\delta_t^{\dot{E}c}$, and $\delta_t^{\dot{E}e}$, for each time step $t \in \mathcal{T}$ with the *uncertainty-scaling factor* ω. E. g., the resulting uncertainty interval for the heating demand is now given by $\left[\max\left\{0, \hat{\dot{E}}_t^h - \omega \cdot \delta_t^{\dot{E}h}\right\}, \hat{\dot{E}}_t^h + \omega \cdot \delta_t^{\dot{E}h}\right]$.

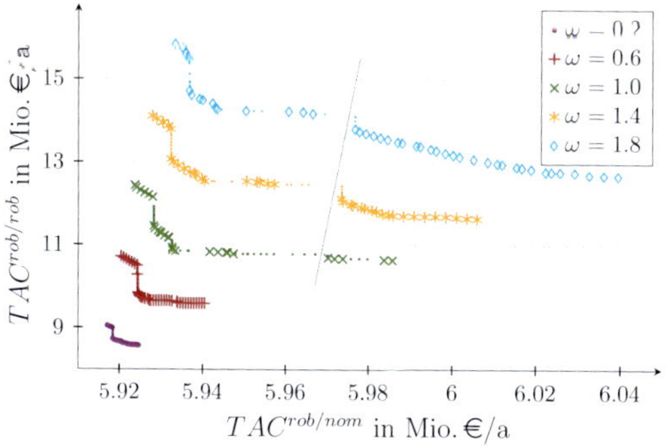

Figure 6.8: Sensitivity of the TRusT curve with respect to the uncertainty-scaling factor ω; solutions for $\omega \geq 1$ contain 3 instead of 2 combined heat and power engines on right-hand side of gray line

We vary the size of the uncertainty significantly ($\omega \in \{0.2, 0.6, 1, 1.4, 1.8\}$). Still, the efficient solutions contain only small deviations in the structure: For instance, 3 instead of 2 combined heat and power engines are contained in the solutions for $\omega \geq 1$ (on the right-hand side of the gray line in Fig. 6.8).

For a high uncertainty-scaling factor ω, the robust design is still cheap for the operation in the nominal scenario: The maximal robust feasible nominal total annualized costs $TAC^{rob/nom}$ are only 2.1 % higher than the nominal optimal total annualized costs of $(TAC^{nom})^* = 5.92$ Mio. €/a. In contrast, robust feasible nominal optimal solutions are expensive for the robust case, especially for large uncertainty-scaling factor ω; e. g., for $\omega = 1.8$, the deviation for different robust designs of the robust costs $TAC^{rob/rob}$ is about 25 % which corresponds to about 3.2 Mio. €/a. Hence, also for the variation of the bounds of the uncertain input parameters, we can derive that the TRusT curve allows for selecting solutions providing good trade-offs.

6.2 Minmax Robust Multi-objective Optimization (RoMO) of distributed energy supply systems

In this section, we employ the concept of RoMO to design robust sustainable energy systems. We show how to transfer this complex mathematical concept introduced by Ehrgott et al. (2014) into practical application for DESS optimization. As most typical case in sustainable design of energy systems, we optimize economic and environmental criteria. Uncertainties are taken into account for energy demands, energy prices, and the specific global warming impact of the electricity mix as introduced in Section 4.2. We reformulate the resulting RoMO problem as an MILP. Thereby, sustainable design problems can be solved using well established methods for multi-objective optimization problems, e. g., ε-constraint method (Mavrotas, 2009). In general, RoMO does no longer lead to a well-defined Pareto front (Ide and Schöbel, 2016). For this reason, it is usually not possible to visualize generated solutions in an easy accessible way. However, we highlight the special case of so-called objective-wise uncertainty leading to an easy way to propose a robust Pareto front. Our investigations show that this special case is most common in the optimization of sustainable energy systems. Hence, our work helps to transfer the general mathematical concept of RoMO into practical engineering problems.

In the case study, we employ RoMO of DESS to investigate the trade-off between economic and environmental criteria for sustainable energy system designs. For better understanding of the robust solutions, we investigate the effect of different types of uncertain parameters.

Major parts of this section are reproduced by permission of Elsevier from:

> Majewski, D. E., Wirtz, M., Lampe, M., Bardow, A. (2017b) Robust multi-objective optimization for sustainable design of distributed energy supply systems. *Computers & Chemical Engineering*, 102:26–39

Contribution report: Principal author, adaption of approach, analysis of results, writing the draft

6.2.1 Applying RoMO to distributed energy supply system design

The concept of RoMO incorporates uncertainties into multi-objective optimization. The mathematical concept was developed and analyzed by Ehrgott et al. (2014). They proposed the robust problem formulation ($\mathcal{MP}_{\mathcal{RC}}$), also called robust counterpart, of a multi-objective problem which enables to find robust efficient solutions for uncertain multi-objective optimization problems:

$$\begin{aligned} (\mathcal{MP}_{\mathcal{RC}}) \quad \min \quad & \sup_{\xi \in \mathcal{U}} f(x,\xi) \\ \text{s.t.} \quad & x \in \mathcal{X}(\xi) \qquad \forall \xi \in \mathcal{U}. \end{aligned}$$

The function $f(x,\xi) = (f_1(x,\xi), \ldots, f_L(x,\xi))^T$ comprises L objective functions $f_i : \mathcal{X} \times \mathcal{U} \to \mathbb{R}$ with $i \in [L]$. We use the notation $[Z] = \{1, \ldots, Z\}$ for sets with any integer $Z \in \mathbb{N}$. Each robust efficient solution x^* minimizes its worst objective function values over all scenarios by minimizing the supremum $\sup_{\xi \in \mathcal{U}} f(x,\xi)$. This inner optimization problem $\sup_{\xi \in \mathcal{U}} f(x,\xi)$ is a multi-objective optimization problem over all scenarios $\xi \in \mathcal{U}$. Hence, the subproblem $\sup_{\xi \in \mathcal{U}} f(x,\xi)$ has no unique, well-defined worst-case scenario but a whole Pareto front of worst-case scenarios (Ide and Schöbel, 2016). As a result, for uncertain multi-objective optimization problems, it is in general not possible to show robust objective function values in an illustrative Pareto front.

However, in optimization of sustainable DESS, a special type of optimization problem enables to still present a robust Pareto front: Most frequently in DESS optimization, the considered objective functions rely on uncertain parameters which are independent, i. e., one uncertain parameter affects at most one objective function. The uncertainty is thus called *objective-wise uncertainty* according to Ehrgott et al. (2014). In our representative case study, uncertainties in the economic objective function have no influence on the environmental objective function and vice versa: Uncertain energy prices do not affect the global warming impact and the total annualized costs do not depend on the uncertain specific global warming impact of purchased energy. The uncertain energy demands influence the objective functions only indirectly. Thus, the uncertainty is objective wise.

Objective-wise uncertainty is the key feature to enable easy visualization of a robust Pareto front, since the complex mathematical concept of RoMO can be simplified: The worst case can be identified for each objective function separately. The resulting robust counterpart for objective-wise ($\mathcal{OW}$) uncertain objective functions is given by Ehrgott et al. (2014):

$$\left(\mathcal{MP}_{\mathcal{RC}}^{\mathcal{OW}}\right) \quad \min \quad \left(\sup_{\xi\in\mathcal{U}} f_1(x,\xi), \ldots, \sup_{\xi\in\mathcal{U}} f_L(x,\xi)\right)^T$$
$$\text{s. t.} \quad x \in \mathcal{X}(\xi) \qquad \forall \xi \in \mathcal{U}.$$

The problem formulation with objective-wise uncertainty leads to a significant advantage: A robust Pareto front can be determined (Ehrgott et al., 2014). For the inner subproblem $\sup_{\xi\in\mathcal{U}} f(x,\xi)$, the worst-case scenario ξ^{wc} leads to one single solution instead of a whole Pareto front, since for each objective function the worst set of parameters can be chosen separately (Example 6.1). Regarding two-stages, x needs to be separated in first-stage variables x^f and second stage variables x^s. For enhanced readability, we shorten the dependency on the DESS model variables in the notation of objective functions and summarize them to x^f and x^s, respectively. Thus, we write $TAC(x^f, x^s, \xi)$ for $TAC\left(\dot{U}, \dot{U}^{el,buy}, \dot{V}^{el,sell}, \gamma, \dot{V}^N, \xi\right)$ and $GWI(x^s, \xi)$ for $GWI\left(\dot{U}, \dot{U}^{el,buy}, \dot{V}^{el,sell}, \xi\right)$, respectively.

Example 6.1 *In Fig. 6.9, a bi-objective minimization problem regarding robust total annualized costs TAC^{rob} and robust global warming impact GWI^{rob} is presented. The set of robust feasible solutions $\mathcal{X} = \left\{(x_1^f, x_1^s), (x_2^f, x_2^s), (x_3^f, x_3^s), (x_4^f, x_4^s)\right\}$ is a discrete set including 4 elements.*

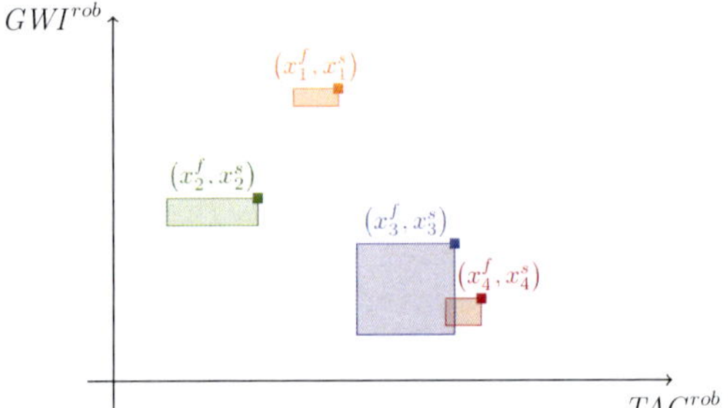

Figure 6.9: Example of RoMO; the range of objective function values of the robust global warming impact GWI^{rob} and the robust total annualized costs TAC^{rob} are presented for the 4 robust feasible solutions $(x^f, x^s) \in \mathcal{X}$. The possible objective function values lie inside large rectangles; small squares highlight the worst-case values $\left(TAC(x^f, x^s, \xi^{wc}),\ GWI(x^s, \xi^{wc})\right)^T$ for each $(x^f, x^s) \in \mathcal{X}$. The solutions (x_2^f, x_2^s), (x_3^f, x_3^s), and (x_4^f, x_4^s) are robust efficient.

Fig. 6.9 shows the ranges of all possible objective function values for each robust feasible solution $(x^f, x^s) \in \mathcal{X}$ *depending on the uncertain scenarios* $\xi \in \mathcal{U}$. *Each range is a rectangle, since each uncertain parameter affects at most one objective function value. Furthermore, prices and the specific global warming impact of electricity can take any value within the fixed intervals given by the uncertainty set* $\mathcal{U}$ *(Eq. (4.6)) leading to continuous ranges. As a result, the Pareto front of the inner subproblem* $\left(\sup_{\xi\in\mathcal{U}} TAC(x^f, x^s, \xi),\ \sup_{\xi\in\mathcal{U}} GWI(x^s, \xi)\right)^T$ *consists of exactly one pair of values* $\left(TAC(x^f, x^s, \xi^{wc}),\ GWI(x^s, \xi^{wc})\right)^T$ *for each robust feasible solution* $(x^f, x^s) \in \mathcal{X}$. *Thus, for the problem* $\left(\mathcal{MP}_{\mathcal{RC}}^{OW}\right)$, *a robust Pareto front is obtained which contains the robust efficient solutions* (x_2^f, x_2^s), (x_3^f, x_3^s), *and* (x_4^f, x_4^s) *(Fig. 6.9). The feasible solution* (x_1^f, x_1^s) *is dominated by* (x_2^f, x_2^s) *and* (x_3^f, x_3^s), *since all possible values for* (x_2^f, x_2^s) *and for* (x_3^f, x_3^s) *are better than the worst possible objective function value for* (x_1^f, x_1^s) *considering both objective functions.*

6.2.2 Problem formulation of RoMO for distributed energy supply systems

For the typical objective functions in the optimization of sustainable DESS, it is possible to visualize robust efficient solutions in a robust Pareto front (Section 6.2.1). However, to find robust efficient solutions, the general RoMO problem formulation needs to be adapted to the optimization of sustainable DESS. In particular, a straightforward application of the mathematical concept of RoMO to DESS optimization cannot be solved directly using common available solvers. We are confronted with two challenges when formulating the RoMO problem which are identical to the challenges related to the TRusT approach (Section 6.1.4): First, the problem is intrinsically infeasible regarding the energy balances. Second, the inner subproblem $\sup_{\xi\in\mathcal{U}} f(x, \xi)$ induces a minmax optimization problem which cannot be solved directly. Hence, a reformulation of the optimization problem is necessary to solve the problem.

To eliminate the intrinsic infeasibility due to the energy balances, we employ the reformulation of energy balances conducted in Section 6.1.4. Hence, we replace equalities by inequalities and only regard the upper bound of the uncertain energy demands (Eq. (6.2)). To reduce overproduction, we additionally enforce the energy balances to hold for uncertain energy demands of the nominal scenario and for the lower bound (Eq. (6.3)).

The second challenge of the reformulation is to eliminate the minmax formulation originating from the supremum over $\xi \in \mathcal{U}$. The supremum over all scenarios needs to be considered when choosing uncertain objective functions, i. e., robust total annualized costs TAC^{rob} or robust global warming impact GWI^{rob}. In RoMO, the robust total annualized costs TAC^{rob} correspond to the robust feasible robust total annualized costs $TAC^{rob/rob}$ as regarded in the TRusT approach (Section 6.1.4). As no nominal costs are regarded in Section 6.2 and, hence, no confusion might occur, the superscript is shortened. In order to formulate an MILP, we proceed in the same way as in Section 6.1.4 and replace the supremum of the uncertain objective functions by auxiliary variables α. The continuous variables α are upper bounds of the uncertain objective functions for all scenarios $\xi \in \mathcal{U}$. By minimizing the upper bounds α, we receive the smallest upper bounds of the uncertain objective functions, i. e., the corresponding supremum. Still, the problem is not solvable because limiting the uncertain objective functions by the auxiliary variables α for *every* scenario $\xi \in \mathcal{U}$ induces an infinite number of constraints. However, in the application of RoMO, it is also sufficient to consider only bounds of the uncertainty set $\mathcal{U}$: For gas prices $\tilde{p}^{gas}$, only the upper limit has to be taken into account. Considering the uncertain electricity prices $\tilde{p}^{el,sell}$ and $\tilde{p}^{el,buy}$, the upper *and* the lower bound need to be considered, since we purchase electricity but also feed in electricity into the grid (Eq. (6.4)). Based on the same reasoning, the uncertain specific global warming impact $\widetilde{GWI}^{el}$ can be replaced by its upper and lower bound as well. Thus, all redundant constraints can be eliminated and only 4 constraints remain to reformulate the objective functions:

$$\begin{aligned}
\min \quad & \begin{pmatrix} \alpha^{TAC} \\ \alpha^{GWI} \end{pmatrix} \\
\text{s.t.} \quad & \sum_{t \in \mathcal{T}} \Bigg[\Delta\tau_t \Bigg(\left(\widehat{p}^{gas} + \delta^{pg} \right) \cdot \sum_{k \in \mathcal{B} \cup \mathcal{CHP}} \dot{U}_{kt} \\
& \qquad + \left(\widehat{p}^{el,buy} + pe \right) \cdot \dot{U}_t^{el,buy} - \left(\widehat{p}^{el,sell} + pe \right) \cdot \dot{V}_t^{el,sell} \Bigg) \Bigg] \\
& \qquad\qquad + \sum_{k \in K} \left(\frac{1}{PVF} + p_k^m \right) \cdot I_k \leq \alpha^{TAC} \\
& \qquad\qquad\qquad pe \in \left\{ -\delta^{pe}, \delta^{pe} \right\} \\
& \sum_{t \in \mathcal{T}} \Delta\tau_t \Bigg[\sum_{k \in \mathcal{B} \cup \mathcal{CHP}} \dot{U}_{kt} \cdot GWI^{gas} \\
& \qquad + \left(\dot{U}_t^{el,buy} - \dot{V}_t^{el,sell} \right) \cdot \left(\widehat{GWI}^{el} + ge \right) \Bigg] \leq \alpha^{GWI} \\
& \qquad\qquad\qquad \forall ge \in \left\{ -\underline{\delta}^{ge}, \overline{\delta}^{ge} \right\}.
\end{aligned} \tag{6.5}$$

The full problem formulation is presented in Appendix C.1. The reformulated robust bi-objective optimization problem for robust sustainable design of DESS is an MILP and thus can be solved employing common available solvers. It is possible to add further uncertain objective functions, e. g., social criteria (Ramos et al., 2014; Mota et al., 2015). For each additional objective function, another auxiliary variable α can be introduced. The additional variable α is minimized while representing the upper bound of the added uncertain objective function for all scenarios $\xi \in \mathcal{U}$, as presented above. If the uncertainty set $\mathcal{U}$ comprises an infinite number of elements, all redundant constraints need to be eliminated, leading to a solvable MILP. The number of newly introduced constraints depends on the number of scenarios which need to be considered, e. g., only 2 additional constraints need to be added when only the upper and lower bound of one set of parameters need to be taken into account at once. Hence, the optimization problems have similar sizes no matter how many objective functions are considered. However, there is a significant increase of computational time to be expected caused by the need of more time to generate the anchor points of the Pareto front. This increase depends on the number of objective functions and not primarily on the uncertainties. Thus, the expected time increase is related to the time increase for solving a deterministic multi-objective optimization problem with an additional objective function.

The proposed extension using auxiliary variables is also applicable to multiple criteria with uncertainties which are not objective-wise. However, the introduction of the auxiliary variables might exclude solutions. A detailed discussion on solving strategies is given by Ehrgott et al. (2014).

6.2.3 Results of the case study

To present the application of RoMO to DESS, the case study presented in Section 4.3 is analyzed and the trade-off between robust total annualized costs TAC^{rob} and robust global warming impact GWI^{rob} is investigated.

All solutions presented in the following sections are computed using the augmented ε-constraint method (Mavrotas, 2009) to guarantee that only efficient solutions are detected. We generate 100 solutions for each Pareto front with GAMS 24.3.3 (McCarl, 2014) using 12.6.0.1 (IBM Corporation, 2015) to solve the resulting optimization problems to machine accuracy. We use a computer with 3.24 GHz and 64 GB RAM employing 4 threads. All values regarding model scales and computational times relate to the problem with the largest superstructure which is solved applying the automated superstructure-generation approach from Voll et al. (2013).

Robust sustainable design options

For generation of the robust Pareto front, we need to consider about 32 300 continuous variables for each solution on the front. The number of binary variables is around 2 100. Each problem comprises about 3 600 constraints. To generate 100 solutions, 20 minutes of computational time are needed.

The effect of uncertain parameters on the Pareto front is assessed in the following. For this purpose, the nominal and three robust Pareto fronts with uncertainties of parameters added consecutively are investigated (Fig. 6.10).

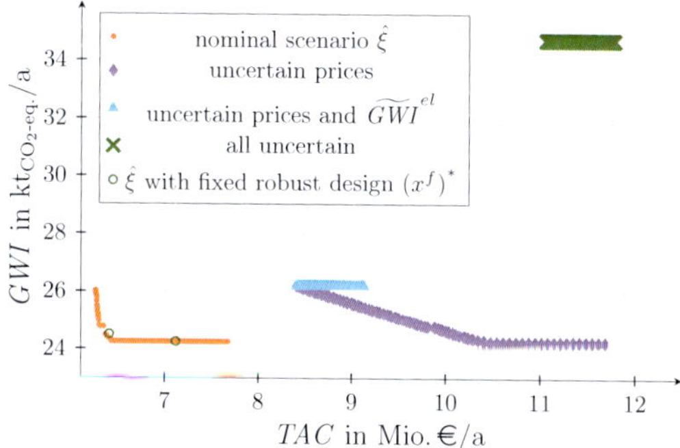

Figure 6.10: Pareto fronts regarding total annualized costs TAC and global warming impact GWI with certain and uncertain parameters: nominal Pareto front for the nominal problem is presented by orange dots; uncertain prices and additionally uncertain specific global warming impact $\widetilde{GWI}^{el}$ are considered for the filled violet rhombuses and for the filled light blue triangles, respectively; adding uncertainty in demands results in the robust Pareto front (dark green crosses); unfilled circles mark solutions for the nominal scenario $\hat{\xi}$ employing the robust optimal design $(x^f)^*$ of robust optimal total annualized costs TAC^{rob} and robust optimal global warming impact GWI^{rob}, respectively.

The uncertain parameters affect the shape of the nominal Pareto front: Uncertain prices stretch the nominal curve along the axis of the total annualized costs and shifts the curve to higher costs. Taking also the uncertainty of the specific global warming impact of the electricity grid $\widetilde{GWI}^{el}$ into account, compresses the former Pareto front obtained with uncertain prices. As a result, no significant trade-off can be observed anymore. Considering all sets of parameters to be uncertain, the uncertain demands shift the Pareto front to a higher global warming impact and to higher total annualized costs. The increase is caused by the higher energy demands which are taken into account.

The shapes of the calculated Pareto fronts differ significantly: For the Pareto front based on perfect foresight, a trade-off between the global warming impact and total annualized costs can be observed. In contrast, there is almost no trade-off for the robust Pareto front taking all uncertainties into account, since the robust global warming impact $GWI^{rob} = 34.7\,\text{kt}_{CO_2\text{-eq.}}/a$ varies only by 0.008 %. The lack of a significant trade-off can be explained by analyzing the investment costs INV. For this purpose, the trade-off between the robust global warming impact GWI^{rob} and the investment costs INV is considered. A detailed discussion can be found in Appendix C.2. For the robust design regarding robust total annualized costs TAC^{rob} and robust global warming impact GWI^{rob}, investment costs lie between 2.8 Mio. € and 4.9 Mio. €. Thus, regarding the robust total annualized costs TAC^{rob} leads to investment costs which can be found on a flat part of the Pareto front regarding investment costs INV and robust global warming impact GWI^{rob} (Fig 6.11).

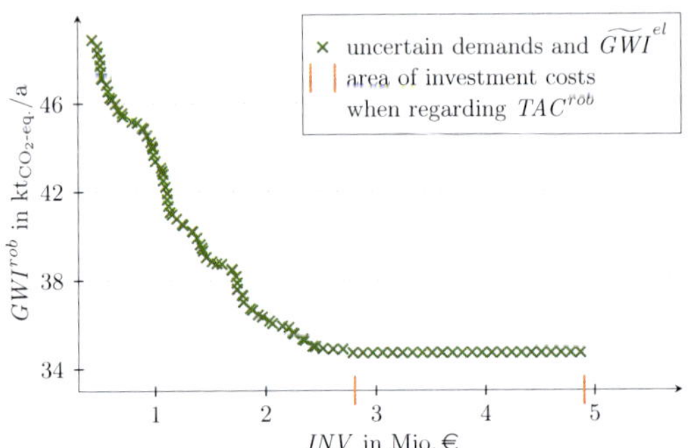

Figure 6.11: Pareto fronts regarding investment costs INV and robust global warming impact GWI^{rob}; since uncertain prices do not affect the uncertain robust global warming impact GWI^{rob} nor investment costs INV, the robust Pareto front only considers uncertain demands and an uncertain specific global warming impact of the electricity mix $\widetilde{GWI}^{el}$; the robust Pareto front is marked by dark green crosses; the area of investment costs INV marked by orange bars refers to the investment costs of the design options when minimizing the robust total annualized costs TAC^{rob} and the robust global warming impact GWI^{rob}

The flat part is caused by the uncertainty of the specific global warming impact of the electricity mix $\widetilde{GWI}^{el}$. In minmax robust optimization, we always consider the worst case. In the case study, the worst case depends on the fact whether electricity is purchased from the grid or fed in: When purchasing electricity, the upper bound of the uncertain specific global warming impact of the electricity mix $\widehat{GWI}^{el} - \overline{\delta}^{ge}$ is employed, whereas the lower bound $\widehat{GWI}^{el} + \underline{\delta}^{ge}$ needs to be regarded when electricity is fed into the grid. This switch in the worst case leads to the fact that the global warming impact cannot be improved by selling electricity to the grid supplied by combined heat and power engines. Only for designs with low investment costs INV, the DESS cannot supply sufficient electricity without purchasing additional electricity from the grid. However, minimizing the robust total annualized costs TAC^{rob} prevents solutions with low investment cost due to high operational costs which would also involve a high robust global warming impact GWI^{rob}. As a result, only a marginal trade-off can be observed. By reducing the specific global warming impact of electricity supplied on site, e. g., by employing renewable energies, a larger trade-off would be expected.

Besides the behavior of solutions in the worst case, we also employ the idea of the TRusT approach and investigate the behavior in the nominal case: The robust optimal designs $(x^f)^*$ of the anchor points from the robust Pareto front perform well for the nominal case. The respective calculated objective function values (6.4 Mio. €, 24.5 $kt_{CO_2\text{-eq.}}/a$) and (7.1 Mio. €, 24.3 $kt_{CO_2\text{-eq.}}/a$) are close to the nominal Pareto front (Fig 6.10, unfilled circles). Thus, robust design options are able to cover all possible demands at small additional costs and small additional environmental impact compared to the nominal case. However, to identify robust optimal designs which perform even better in the nominal scenario, RoMO is combined with the TRusT approach in Section 6.3.

The robust efficient solutions (Fig. 6.12b) show similar characteristics as the nominal efficient solutions (Fig. 6.12a): Combined heat and power engines and absorption chillers replace boilers and compression chillers when the robust global warming impact GWI^{rob} decrease. Before increasing the number of combined heat and power engines from 2 components to 3 components a newly installed boiler is used in one design to increase the available heating for the absorption chillers. The characteristics of the solutions reflect the fact that the specific global warming impact of trigeneration systems is lower than covering the demands separately with boilers, compression chillers, and the electricity grid. However, combined heat and power engines are more expensive in terms of investment costs. Regarding the robust efficient solutions, the uncertain specific global warming impact of the electricity mix $\widetilde{GWI}^{el}$ leads to significant differences in the design: In total, the robust efficient solutions contain less thermal power of combined heat and power engines than the nominal efficient solutions. The reason is that the connection to the electricity grid is not advantageous when regarding an uncertain specific global warming impact $\widetilde{GWI}^{el}$, since a positive effect on the global warming impact by selling electricity cannot be guaranteed.

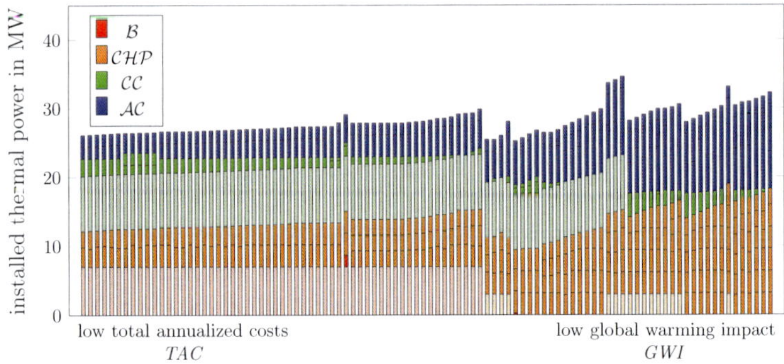

(a) Efficient design (structure and sizing) of installed components.

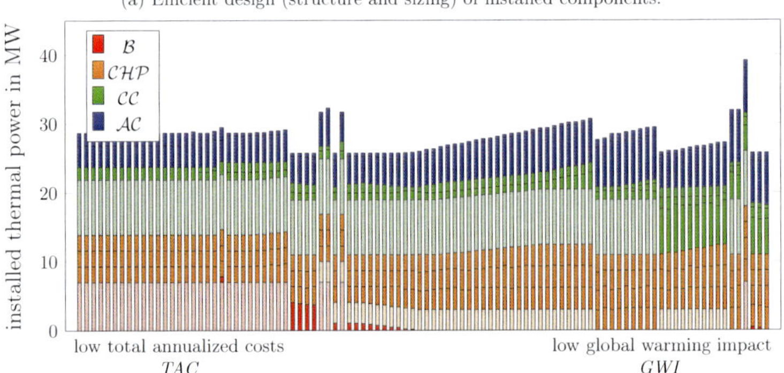

(b) Robust efficient design (structure and sizing) of installed components.

Figure 6.12: $\mathcal{B}$ boilers, $\mathcal{CHP}$ combined heat and power engines, $\mathcal{CC}$ compression chiller, and $\mathcal{AC}$ absorption chillers; components shown in light colors remain from the already existing components.

6.3 Combination of the TRusT approach and RoMO

To design sustainable DESS while regarding the trade-off between nominal and worst-case costs, the TRusT approach is combined with RoMO. By the combination, robust sustainable design options are also assessed for the nominal scenario.

6.3.1 Problem formulation of the combination of the TRusT approach and RoMO

The combination of the TRusT approach with RoMO is easy to conduct: Since both approaches depend on the same constraints (for an overview see Appendix C.1), only the objective functions needs to be unified. For this purpose, we consider three instead of two objective functions, i. e., the nominal total annualized costs $TAC^{rob/nom}$, the robust total annualized costs TAC^{rob}, and the robust global warming impact GWI^{rob}. To determine the robust values, the suprema need to be considered; thus, for the optimization problem, auxiliary variables α are needed (Section 6.1.4 and 6.2.2). The auxiliary variables α limit the suprema of the objective functions (Eq. (6.5)):

$$\min \begin{pmatrix} TAC^{rob/nom} \\ \alpha^{TAC} \\ \alpha^{GWI} \end{pmatrix}.$$

6.3.2 Results of the case study

The Pareto fronts are computed employing the augmented ε-constraint method (Mavrotas, 2009). To obtain an equally distributed Pareto front, the problem is solved 3 times: Each objective function is minimized while the remaining two objective functions are limited. Calculating the Pareto front with about 200 points takes 63 minutes. We solve the problem to machine accuracy with GAMS 24.7.3 (McCarl and Rosenthal, 2016) using CPLEX 12.6.3.0 (IBM Corporation, 2015). A computer with 3.24 GHz and 64 GB RAM employing 4 threads is used.

In Fig. 6.13, the Pareto front of the combination of the TRusT approach and RoMO is shown regarding the nominal total annualized costs $TAC^{rob/nom}$, the robust total annualized costs TAC^{rob}, and the robust global warming impact GWI^{rob}.

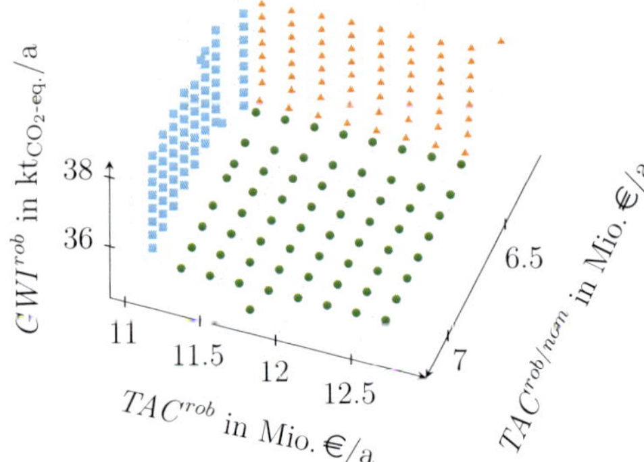

Figure 6.13: 3-dimensional Pareto front regarding the nominal total annualized costs $TAC^{rob/nom}$, the robust total annualized costs TAC^{rob}, and the robust global warming impact GWI^{rob}; orange triangles are computed minimizing the nominal total annualized costs $TAC^{rob/nom}$, blue squares minimizing the robust total annualized costs TAC^{rob}, and green dots minimizing the robust global warming impact GWI^{rob}; for improved visualization of the Pareto front showing 3 different perspectives see Fig. C.6

Remarkable is the shape of the computed Pareto front. Each part of the Pareto front calculated by minimizing one criterion while limiting the other two criteria is very steep (marked by orange triangles, blue squares, and green dots in Fig. 6.13). As a result, the Pareto front resembles 3 sides of a cube with edges rounded off. Limiting the nominal total annualized costs $TAC^{rob/nom}$ besides the robust total annualized costs TAC^{rob} when minimizing the robust global warming impact GWI^{rob} leads to a significant trade-off. The variation of the robust global warming impact GWI^{rob} is about 10.0 %. As comparison, when regarding only the robust total annualized costs TAC^{rob} and the robust global warming impact GWI^{rob}, no significant trade-off can be observed (Section 6.2.3). When a reduction in the nominal total annualized costs $TAC^{rob/nom}$ is aspired, a high increase in the robust global warming impact GWI^{rob} has to be accepted. When comparing the results with the results of the TRusT approach (Section 6.1.5), the same observation holds for the nominal total annualized costs $TAC^{rob/nom}$: When a third objective function is considered, i. e., the global warming impact GWI^{rob}, the variation of the nominal total annualized costs $TAC^{rob/nom}$ increases significantly. Precisely, the variation increases from only 1 % on the TRusT curve (Section 6.1.5) up to 13.4 %.

The calculated design options show only small changes compared to the design options of RoMO (Fig. 6.12b). Fig. 6.14 shows design options for minimized robust global warming impact GWI^{rob} with limited robust total annualized costs TAC^{rob} and limited nominal total annualized costs $TAC^{rob/nom}$ (green dots in Fig. 6.13). Fig. 6.14a shows the already observed characteristics (Section 6.2.3). For decreasing robust global warming impact GWI^{rob}, trigeneration components are enlarged. Higher thermal power of combined heat and power engines as well as a higher thermal power of absorption chillers is installed, whereas the thermal power of installed compression chillers and boilers is reduced (Fig. 6.14a).

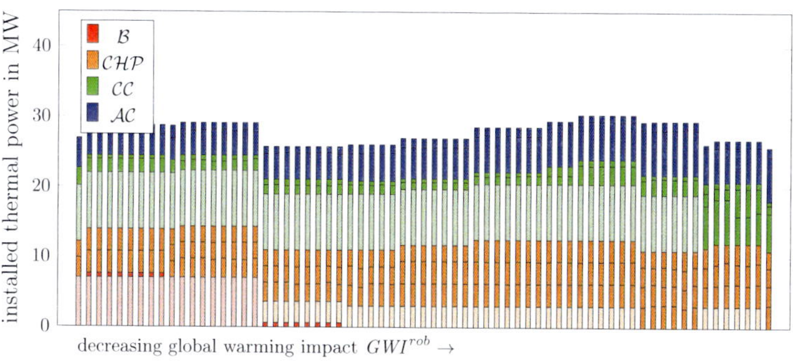

(a) Robust efficient design (structure and sizing) of installed components sorted by decreasing robust global warming impact GWI^{rob}

Figure 6.14: Design of efficient solutions computed while minimizing robust global warming impact GWI^{rob} (green dots in Fig. 6.13); 3 modes of sorting are presented: sorted by (a) decreasing robust global warming impact GWI^{rob}, by (b) increasing robust total annualized costs $TAC^{rob/nom}$, and by (c) increasing nominal total annualized costs TAC^{rob}; $\mathcal{B}$ boilers, $\mathcal{CHP}$ combined heat and power engines, $\mathcal{CC}$ compression chiller, and $\mathcal{AC}$ absorption chillers; components shown in light colors remain from the already existing components

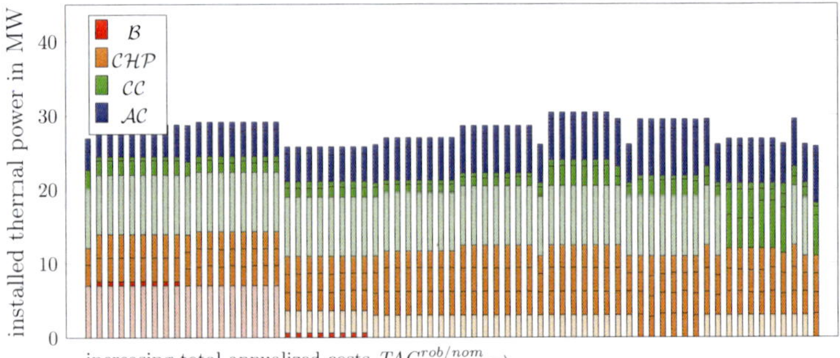

(b) Robust efficient design (structure and sizing) of installed components sorted by increasing robust total annualized costs $TAC^{rob/nom}$

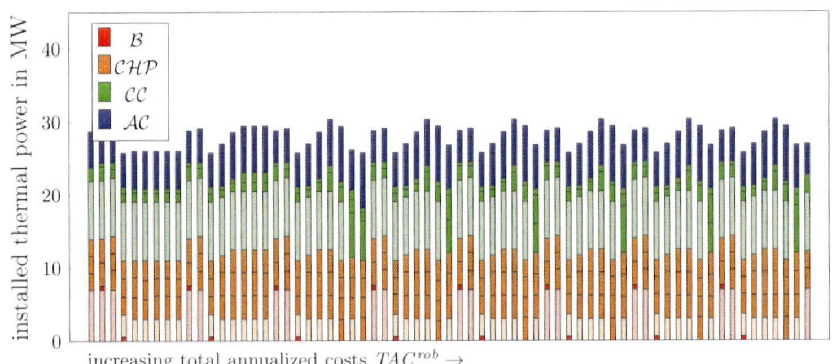

(c) Robust efficient design (structure and sizing) of installed components sorted by increasing nominal total annualized costs TAC^{rob}

The different sorting of the design options allows deeper insight of the characteristics of solutions marked by green dots in Fig. 6.13. The solutions marked by green dots resemble a grid with lines of nearly constant values for the nominal total annualized costs $TAC^{rob/nom}$ and the robust total annualized costs TAC^{rob}, respectively. Fig. 6.14b corresponds to sorting the solutions line by line with respective stable robust total annualized costs TAC^{rob} and increasing nominal total annualized costs $TAC^{rob/nom}$; Fig. 6.14c corresponds to sorting the solutions line by line with respective stable nominal total annualized costs $TAC^{rob/nom}$ and increasing robust total annualized costs TAC^{rob}.

When the designs are sorted by increasing nominal total annualized costs $TAC^{rob/nom}$ (Fig. 6.14b), the designs follow a clear trend: For increasing nominal total annualized costs $TAC^{rob/nom}$, the thermal power of newly installed components increases and the thermal power of already existing components reduces. The reason for the trend is that allowing higher nominal total annualized costs $TAC^{rob/nom}$ enables increasing investment costs of the system design which also enables a smaller robust global warming impact GWI^{rob}.

Regarding the designs sorted by increasing robust total annualized costs TAC^{rob} (Fig. 6.14c), no trend of the design options can be observed. The varying capacities and thus varying investment costs reveal that the investment costs are less decisive than operational costs. Since the robust total annualized costs TAC^{rob} take into account high demands, the operational costs become more important and the influence of the investment costs on the robust total annualized costs TAC^{rob} decreases. However, the effect of increasing nominal total annualized costs $TAC^{rob/nom}$ along each line with respective stable robust total annualized costs TAC^{rob} is still visible, apparent by similar designs repeating after about 8 designs.

The designs of the other parts of the Pareto front (solutions marked by orange triangles and blue squares in Fig. 6.13) do not show remarkable trends (Appendix C.7).

6.4 Conclusions

In the following, summaries and conclusions are given for the approaches incorporating uncertainty of input parameters into the framework.

6.4.1 Conclusions of the TRusT approach

We show how robust design of energy supply systems can provide both, cost-efficiency solutions *and* a sufficient energy supply. For this reason, we propose the TRusT approach to integrate uncertainty of input parameters into the design of DESS: The bi-objective approach minimizes both, the *nominal* and the *minmax robust* objective function. DESS optimization can be interpreted as a two-stage problem as discussed in Section 2.2. In the TRusT approach, we additionally include uncertainty of input parameters. We incorporate the two-stage nature of DESS in the TRusT approach: Second-stage variables (operational variables) are determined separately for the nominal and the robust objective function, while the robust first-stage variables (design variables) are shared. A *robust design* has to satisfy all uncertain demands to ensure sufficient energy supply.

The TRusT approach can be applied easily for interval-based uncertainty, as the TRusT problem can be reformulated with low effort as bi-objective MILP. The resulting *TRusT curve* represents the Pareto front of the nominal (i. e., expected) and robust objective functions employing identical robust first-stage variables. Thus, we obtain information about how the same robust design performs in everyday business and in the worst case.

In the case study, uncertain demands and prices are considered while minimizing the total annualized costs. Applying the TRusT approach shows that implementing a minmax robust design implies only low additional costs for the nominal operation. The maximal nominal costs employing a robust design lie only 1.2 % above the single-objective nominal optimal costs. In contrast, the robust total annualized costs turn out to be high for a nominal optimal robust design if the parameters differ from the nominal scenario. The sensitivity study shows that even for high uncertainties the minmax robust design is still cost-efficient for the nominal scenario.

6.4.2 Conclusions of RoMO and of the combination with the TRusT approach

Sustainable design of DESS inherently involves multiple objectives. As discussed in Section 2.3, economic and environmental criteria are typically employed to design sustainable energy systems. However, the input parameters of the optimization problem are usually uncertain in real life. This uncertainty needs to be taken into account. For this reason, we transfer the mathematical concept of RoMO, as introduced by Ehrgott et al. (2014) to the problem of designing a sustainable DESS. We propose a RoMO problem formulation to determine robust sustainable design options coping with uncertainties. Reformulations lead to an MILP allowing to apply established solvers and methods for multi-objective optimization, e. g., the ε-constraint method.

In the real-world case study, we consider the total annualized costs and the global warming impact. We assume uncertainties for the demands, energy prices, and the specific global warming impact of the future electricity mix as introduced in Section 4.2.

Our analysis investigates the effect of the different uncertain parameters on the shape of the Pareto front: Uncertainty of prices lead to higher costs with an increased cost range. The uncertainty of the global warming impact decreases the range of the global warming impact of the design options and thus reduces the trade-off. The effect is caused by the system aspiring to be autonomous of the electricity grid once sufficient thermal power is installed. Uncertain energy demands shift the Pareto front to higher values regarding both objective functions.

Employing the idea of the TRusT approach, robust efficient design options of the anchor points identified by RoMO are assessed in the nominal scenario. The analysis shows that robust design options also perform remarkably well for the nominal scenario assuming parameters to be known with perfect foresight. However, to assess the whole robust Pareto front, we combine the TRusT approach with RoMO employing 3 objective functions: the nominal total annualized costs, the robust total annualized costs, and the robust global warming impact. The results show remarkable trade-offs between the considered criteria. However, the designs do not change significant for the respective minimized objective function computed by the augmented ε-constraint method. Only for minimized robust global warming impact, slightly changes related to higher investments in new components are apparent.

6.4.3 Conclusions of robustness against uncertain input parameters

The identified robust design options by the TRusT approach cover uncertain demands and guarantee sufficient energy supply for the considered scenarios. Our investigations show that sufficient and cost-efficient energy supply are not mutually exclusive, but can be integrated using the TRusT approach leading to robust and cost-efficient designs of DESS.

Applying the proposed formulation for RoMO enables the decision maker to find promising solutions for engineering practice. The identified design options for DESS are robust as well as well-performing regarding sustainability. The proposed approaches are easy to apply and the resulting optimization problem can be solved in reasonable time. As a result, the approaches allow to include robustness into the framework in an easy and effective way.

Both approaches can be combined to regard the trade-off between the nominal costs, the worst-case costs, and the worst environmental impact (Section 6.3). However, RoMO already identifies design options which perform well in the nominal case, even though this cannot be assumed in general.

Applying either the TRusT approach or RoMO computes multiple design options. Hence, the decision maker still needs to select one design for implementation.

Chapter 7

The flexible here-and-now decision (flex-hand) approach for design selection in multi-objective optimization

The selection of an energy system design considering multiple criteria is complex, since the Pareto front usually contains a diverse set of design options. To help the decision maker implementing one fixed design, we propose the ***f**lexible **h**ere-**an**d-now **d**ecision (flex-hand)* approach. The approach selects one single design which represents the whole Pareto front best without depending on any additional information of the decision maker. Integration of the flex-hand approach into the framework of this thesis enables selecting the best sustainable design (Fig. 7.1). The best sustainable design allows flexible operation regarding the employed sustainability criteria during optimization.

Major parts of this section are reproduced from:

> Hollermann, D.E., Goerigk, M., Hoffrogge, D.F., Hennen, M., Bardow, A. (2019a) Flexible here-and-now decisions for two-stage multi-objective optimization: Method and application to energy system design selection. *arXiv e-prints*, arXiv:1906.08621v1

Contribution report: Principal author, enhancement and further development of the approach based on a first proposition by Marc Goerigk, implementation of the approach, analysis of results, writing the draft

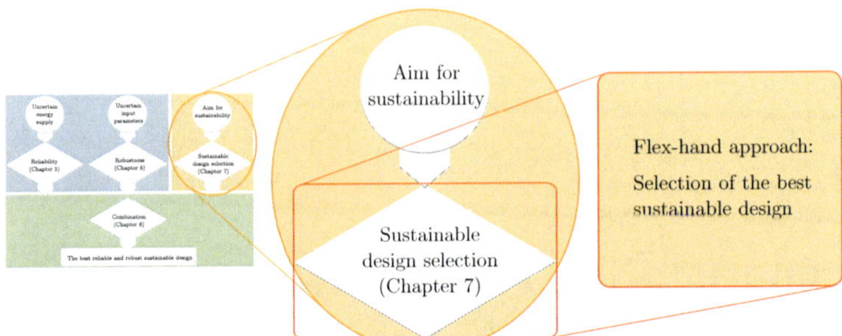

Figure 7.1: Overview Chapter 7 – flex-hand approach to automatically select one single design

The flex-hand approach automatically selects the best design looking even beyond the set of already identified efficient solutions. For this purpose, we minimize the distance between the Pareto front of the synthesis problem, i. e., the Pareto front with changing design options on the first stage, and the Pareto front induced by one fixed design. The fixed design with the minimal distance is selected by the proposed approach, since this design provides high flexibility regarding the considered objective functions. Thus, the second stage can be well adapted to a changing focus from one to another objective function.

For two-stage problems, our approach allows automatic selection of one design regarding multiple decision criteria. While we introduce our approach in the context of DESS, the methodology is general and can be applied to any two-stage multi-objective optimization problem.

Before presenting the flex-hand approach in Section 7.2 and the computational results in Section 7.3, we first introduce some basic concepts and notation in Section 7.1. We summarize and conclude Chapter 7 in Section 7.4.

7.1 Two-stage multi-objective optimization

The flex-hand approach employs multiple objectives to design sustainable DESS. As objectives might be conflicting, we are interested in trade-off solutions. A solution is called efficient if there is no other solution that is at least as good in each objective and strictly better in at least one objective (Section 2.3). For a two-stage problem as DESS optimization, the two-stage multi-objective problem formulation with L objectives is given by:

$$\begin{aligned} \min \; & \Big(f_1(x^f, x^s), \ldots, f_L(x^f, x^s) \Big)^T \\ \text{s.t.} \; & x^f \in \mathcal{X}^f \\ & x^s \in \mathcal{X}^s(x^f)\,, \end{aligned}$$

with $\mathcal{X}^f$ being the set of feasible solutions for first-stage variables x^f, and $\mathcal{X}^s(x^f)$ being the set of feasible solutions for the second-stage variables x^s which depends on the chosen first-stage solution x^f. The corresponding Pareto front, is denoted by $\mathcal{P}^* = \{\rho^*(1), \ldots, \rho^*(N)\}$ with $\rho^*(j) = \Big(\rho_1^*(j), \ldots, \rho_L^*(j)\Big)^T \in \mathbb{R}^L$ for $j \in [N]$. In this thesis, we consider a discrete subset of the Pareto front A discrete representative set of efficient solutions can be obtained, e. g., by using the ε-constraint method (Ehrgott, 2005). For upper and lower bounds of the ε, the ideal and the nadir point could be chosen limiting the possible range of objective functions. If the ideal and the nadir point do not exist for the regarded problem, bounds have to be defined manually by the decision maker (Miettinen, 2008) or an a priori approach has to be employed to restrict the regarded region of interest, e. g., by finding knee-regions of the Pareto front (see Section 2.3).

We call the problem *ideal* when both first-stage variables x^f and second-stage variables x^s can be chosen separately for each efficient point. The corresponding set of efficient solutions in the objective space $\mathcal{P}^*$ is called *ideal Pareto front*. The word *ideal* emphasizes that the ideal Pareto front is always better than the Pareto front with fixed first-stage variables. In DESS optimization, the ideal Pareto front would imply changing design options (e. g. heating equipment) along the Pareto front and thus, cannot be reached by DESS implemented in the real world. In our flex-hand approach, we use the ideal Pareto front as benchmark for evaluating Pareto fronts with fixed design options, i. e., with fixed first-stage variables.

7.2 The flex-hand approach

The goal of the flex-hand approach is to find one fixed design which represents the ideal Pareto front of the synthesis problem best. In optimization of DESS, the *operation* of an installed system can be adapted but changing the installed system *design* is not possible in the short term. The flex-hand approach selects as best design the design with the highest adaptability to the Pareto front of the multi-objective optimization problem regarding the operation. To determine such a highly flexible design, the flex-hand approach minimizes the distance between the ideal Pareto front and the Pareto front based on one fixed design but flexible operation.

The flex hand approach is not limited to DESS optimization but can be applied to any two-stage multi-objective problem where the first-stage variables x^f need to be fixed right now and the second-stage variables x^s can be determined later.

For one fixed design, i. e., for a given first-stage solution x^f, we calculate efficient solutions:

$$\begin{aligned}&\min \left(f_1(x^f, x^s), \dots, f_L(x^f, x^s)\right)^T \\ &\text{s.t. } x^s \in \mathcal{X}^s(x^f)\,.\end{aligned}$$

We obtain a Pareto front $\mathcal{P}(x^f) = \left\{p(x^f, 1), \dots, p(x^f, N(x^f))\right\}$ depending on the first-stage solution x^f with $N(x^f)$ points which we call *fixed first-stage Pareto front.*

Now the question arises: How to choose first-stage variables x^f such that we get the best design? To determine the quality of a first-stage solution x^f, we compare the Pareto front $\mathcal{P}(x^f)$ with fixed first-stage to the ideal Pareto front $\mathcal{P}^*$. For the comparison of two sets of multi-objective solutions, a variety of distance measures have been developed (for a review see Zitzler et al. (2003)). In this thesis, we choose a comparison metric based on an additive *binary ε-indicator*. As discussed by Zitzler et al. (2003), there is no single best way to compare Pareto fronts, but the ε-indicator is proposed as a good overall method. Still, employing other metrics would be possible in our setting.

For two sets $\mathcal{P}^1 = \{p(1), \dots, p(S)\}$ and $\mathcal{P}^2 = \{q(1), \dots, q(T)\}$ in the L-dimensional objective space, the binary ε-indicator is obtained by

$$D(\mathcal{P}^1, \mathcal{P}^2) = \min\left\{\varepsilon \;:\; \forall l \in [T]\; \exists j \in [S] \text{ s.t. } p_i(j) - q_i(l) \le \varepsilon \;\forall i \in [L]\right\}. \quad (7.1)$$

For our approach, the measure $D\left(\mathcal{P}(x^f), \mathcal{P}^*\right)$ indicates the distance between a fixed first-stage Pareto front and the ideal Pareto front.

The comparison metric can be interpreted as follows: Recall that each point of the ideal Pareto front $\mathcal{P}^*$ may involve changing first-stage solutions x^f along the front. In contrast, the fixed first-stage Pareto front $\mathcal{P}(x^f)$ is based on one single first-stage solution x^f where only the second-stage decision can be adapted. Figure

7.2 shows the comparison of the ideal Pareto front to an arbitrary fixed first-stage Pareto front.

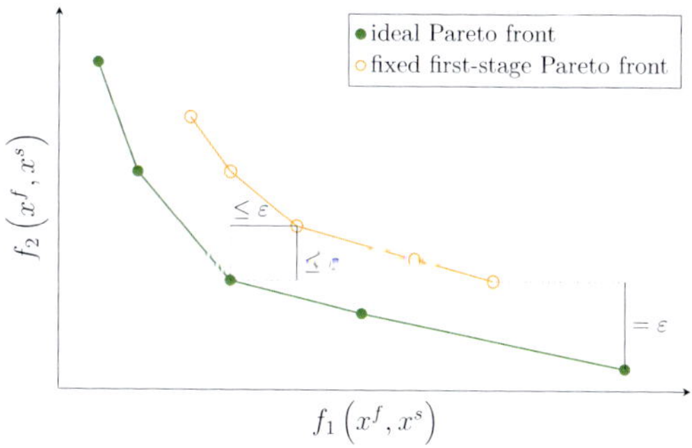

Figure 7.2: Comparison of ideal Pareto front to an arbitrary fixed first-stage Pareto front to assess the quality of the fixed first-stage solutions x^f; dark green dots: ideal Pareto front used as benchmark; orange circles: fixed first-stage Pareto front with distance ε to the ideal Pareto front; lines are included to guide reader's eyes

For each point on the ideal Pareto front, a point on the fixed first-stage Pareto front can be determined such that the difference in each objective function is smaller than a value of ε. By minimizing ε, the distance between the Pareto fronts is minimized. In this way, for DESS, we minimize the distance between the ideal Pareto front allowing changing designs as well as operation and the Pareto front based on one fixed design where only the operation can be adapted.

Since we consider the difference regarding each objective function separately, we normalize the objective functions based on the value range in $\mathcal{P}^*$ to circumvent misleading effects by different scales of objective values. Towards this end, we use the normalized objectives $\overline{f}_i$ with

$$\overline{f}_i(x^f, x^s) = \frac{f_i(x^f, x^s) - \min_{j\in[N]} \rho_i^*(j)}{\max_{j\in[N]} \rho_i^*(j) - \min_{j\in[N]} \rho_i^*(j)} . \tag{7.2}$$

In the following, we write $\overline{\mathcal{P}}(x^f)$, $\overline{\mathcal{P}}^*$ to denote normalized Pareto fronts.

Searching for an optimal first-stage variable $(x^f)^*$, i. e., an optimal design, we now minimize the distance between the Pareto fronts:

$$\min\left\{D\left(\overline{\mathcal{P}}(x^f),\overline{\mathcal{P}}^*\right) : x^f \in \mathcal{X}^f\right\}.$$

The problem formulation can also be written as:

$$\begin{aligned}
\min\ & \overline{\varepsilon} & \\
\text{s.t.}\ & \overline{f}_i(x^f, x^s(j)) - \overline{p}_i^*(j) \leq \overline{\varepsilon} & \forall i \in [L], j \subset [N] \\
& x^f \in \mathcal{X}^f & \\
& x^s(j) \in \mathcal{X}^s(x^f) & \forall j \in [N],
\end{aligned}$$

where $\overline{p}_i^*(j)$ represents the normalized objective function value of point j on the ideal Pareto front $\overline{\mathcal{P}}^*$. For each point j on the ideal Pareto front $\overline{\mathcal{P}}^*$, we consider one point on the fixed first-stage Pareto front $\overline{\mathcal{P}}(x^f)$. Note that given a first-stage solution x^f, the number of points $N(x^f)$ of the resulting fixed first-stage Pareto front $\overline{\mathcal{P}}(x^f)$ might be different from the number N of points of the ideal Pareto front $\overline{\mathcal{P}}^*$. However, for the purpose of identifying an optimal first-stage solution $(x^f)^*$, it is sufficient to consider exactly N points on the fixed first-stage Pareto front $\overline{\mathcal{P}}(x^f)$. If $N(x^f) > N$ holds, we only need to identify one point on the fixed first-stage Pareto front $\overline{\mathcal{P}}(x^f)$ for each point on the ideal Pareto front $\overline{\mathcal{P}}^*$ which minimizes the distance; recall, for every $l \in [T]$ there needs to exist one $j \in [S]$ as specified by the definition of the binary ε-indicator (Eq. (7.1)). If $N(x^f) < N$ holds, we can duplicate some second-stage solutions x^s to obtain N points without changing the respective objective value. The complete flex-hand model for DESS optimization is presented in Appendix D.1.

The flex-hand approach yields an optimal first-stage solution $(x^f)^*$ which represents the ideal Pareto front best regarding the chosen measure. We call the optimal first-stage solution $(x^f)^*$ the *flex-hand solution*. In general, the flex-hand solution chosen by our approach is not necessarily part of the solutions of the calculated ideal Pareto front $\mathcal{P}^*$. Thus, approaches based on sorting or solution-reduction of the ideal Pareto front would not be able to identify the flex-hand solution in general.

Having found a flex-hand solution, we calculate the corresponding efficient second-stage solution x^s in a separate post-processing optimization step. The resulting Pareto front is called the *flex-hand Pareto front*. The number of points on the flex-hand Pareto front might differ from the original number of points N. For DESS, this post-processing optimization corresponds to an operational multi-objective optimization based on a fixed design.

Applying the flex-hand approach to single-stage optimization problems is also possible. For single-stage problems, the flex-hand approach reduces to an approach minimizing the objective-wise distance to the ideal point. As there is no second stage, the distance is measured only from a point instead of a whole Pareto front to the ideal Pareto Front. Hence, only the anchor points are decisive when determining the minimal distance. As we measure the distance objective-wise, the distance is equivalent to the distance to the ideal point (Appendix D.2).

There is also an alternative interpretation of our proposed flex-hand approach: To solve two-stage multi-objective optimization problems, weighted sums such as $\sum_{i\in[L]} \lambda_i f_i(x^f, x^s)$ could be considered (Ehrgott, 2005). Since we do not know the preferred preference weighting, the weighting factors λ_i are uncertain. In this interpretation, each point on the ideal Pareto front represents the optimal solutions if we knew the preference weighting in advance. The aim is now to find a first-stage solution x^f which yields a high solution quality for a wide range of weights. The $\bar{\varepsilon}$ value of our flex-hand approach can thus be considered as regret (see Aissi et al. (2009) for a survey of regret optimization). However, employing weighted sums, only solutions on the convex hull of the Pareto front can be found. In contrast, our approach also considers points not on the convex hull of Pareto solutions.

In Appendix D.3, a scenario-based flex-hand approach is provided to incorporate uncertainties into the flex-hand approach. However, the scenario-based approach can only cope with a discrete uncertainty set $\mathcal{U}$. Apart from this limitation, previous investigations of Mayer (2018) have shown that the combination of RoMO with the flex-hand approach yields Pareto fronts which partially dominate the scenario-based flex-hand Pareto front when evaluated for the discrete scenarios. The results are based on an extended version of the case study employed in this thesis and regards also renewable energies (Baumgärtner et al., 2019b). Thus, in Chapter 8, we employ RoMO instead of the scenario-based flex-hand approach in our framework of this thesis.

7.3 Results of the case study

In this section, the flex-hand approach is applied to design the energy system of a real-world industrial park (Section 4.3). For the design optimization, we assume a "green field" without existing energy conversion components. However, the flex-hand approach could also be applied to retrofit a DESS. In this section, the price for purchasing gas is assumed to be $\widehat{p}^{gas} = 6\,\text{ct/kWh}$. To compute Pareto fronts, we use the adaptive normal boundary intersection method (Das and Dennis, 1998). For calculation, we employ 4 threads of a computer with 3.24 GHz and 64 GB RAM. The problem is formulated in GAMS 24.7.3 (McCarl and Rosenthal, 2016) and solved by the solver CPLEX 12.6.3.0 (IBM Corporation, 2015) to machine accuracy. All values regarding model scales and computational times relate to the problem with the largest superstructure which is solved applying the automated superstructure-generation approach from Voll et al. (2013).

The flex-hand design

We now employ the flex-hand approach to design the sustainable DESS in order to obtain the best solution for the first-stage variables x^f which we call the *flex-hand design*. For this purpose, we first calculate the ideal Pareto front as benchmark. The ideal Pareto front is obtained by allowing a different design for each point on the front. The largest optimization problem for calculating a point on the ideal Pareto front consists of 1 950 equations, 576 variables, and 310 binary variables after presolve. In total, the whole ideal Pareto front is calculated in 5 minutes and 17 seconds. The flex-hand problem has 5 099 equations, 2 023 variables, and 751 binary variables after presolve. Computing the flex-hand Pareto front takes 2 minutes and 32 seconds. Both the ideal Pareto front and the flex-hand Pareto front are shown in Fig. 7.3.

The flex-hand Pareto front of the selected design is "stretched out". Thus, the flex-hand design allows for flexible operation providing a high ability to adapt to changing future objectives. The flex-hand design can be operated such that the total annualized costs TAC are very low at 7.78 Mio.€/a or such that the global warming impact GWI is very low with $22.60\,\text{kt}_{\text{CO}_2\text{-eq.}}/\text{a}$.

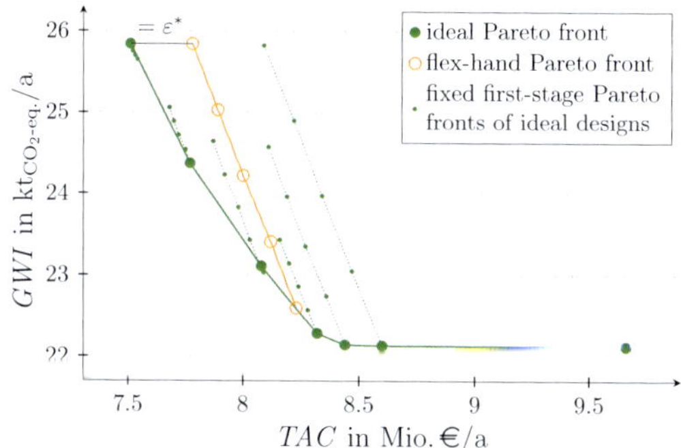

Figure 7.3: Comparison of ideal Pareto front (dark green dots) and the flex-hand Pareto front (orange circles) with minimal distance ε^* to the ideal Pareto front; small dark green dots: fixed first-stage Pareto front of ideal designs, i. e., efficient operation for each design of the ideal Pareto front; Pareto fronts are presented without normalization; lines are included to guide reader's eyes

In this case study, the minimal distance ε^* between the ideal and the flex-hand Pareto front is limited, e. g., by the anchor points with minimal total annualized costs limits (Fig. 7.3). The corresponding scaled value for the minimized distance is $\bar{\varepsilon}^* = 0.128$. For unscaled values, the minimal total annualized costs for the ideal design are 7.51 Mio.€/a and for the flex-hand design 7.78 Mio.€/a. Hence, the maximal deviation for total annualized costs is 0.27 Mio.€/a which corresponds to a maximal loss of only 3.60 % compared to the ideal design with minimized total annualized costs. Regarding the minimal global warming impact, the maximal deviation is only 2.17 %. Thus, the flexibility of the flex-hand design is very high regarding both objective functions.

When having a closer look at the selected design, we cannot identify a single reason for its higher flexibility in operation (Fig. 7.4).

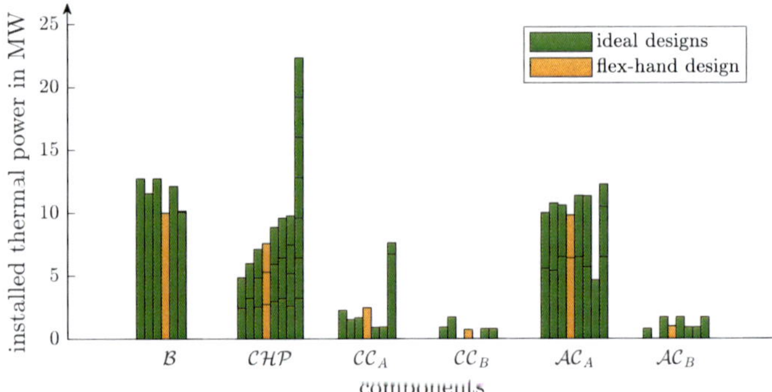

Figure 7.4: In dark green: designs of ideal Pareto front; in orange: flex-hand design of the flex-hand Pareto front; from left to right for each component type, the designs are ordered by decreasing minimal global warming impact; $\mathcal{B}$ boiler, $\mathcal{CHP}$ combined heat and power engine, $\mathcal{CC}_A$ and $\mathcal{CC}_B$ compression chillers, and $\mathcal{AC}_A$ and $\mathcal{AC}_B$ absorption chillers installed on Site A and Site B, respectively

In general, solutions with lower global warming impact prefer installing higher thermal power of combined heat and power engines, since the specific global warming impact of the electricity mix of the grid is higher than the impact of the combined heat and power engines in combination with absorption chillers. The flex-hand design does not show remarkable differences compared to the other ideal designs but provides an excellent compromise. Without the proposed approach, this highly adaptable design would most likely not have been selected by the decision maker.

7.4 Conclusions

The sustainable optimization of DESS is inherently a two-stage optimization problem with multiple decision criteria. Applying multi-objective optimization usually generates different designs for each point on the Pareto front. We propose the flex-hand approach to select one single design which performs well regarding all decision criteria. The idea of the flex-hand approach is to approximate the Pareto front with changing design options (*ideal Pareto front*) by a Pareto front with one fixed design for the whole front. The design leading to the minimal distance between both Pareto fronts is the design which is selected by the flex-hand approach. The selected design (*flex-hand design*) is able to adapt well to all regarded criteria and, thus, provides high flexibility to reach future aims for which focus might change between the considered objective functions.

The real-world case study demonstrates the resulting high adaptability with respect to the considered criteria. The calculated Pareto front of the flex-hand design is "stretched out" in comparison to the Pareto fronts obtained by operational optimization of other designs lying on the ideal Pareto front. The objective function values of flex-hand design differ by less than 3.6 % from the ideal values which highlights the excellent quality of the selected flex-hand design. The flex-hand design does not show remarkable differences compared to the designs lying on the ideal Pareto front. Thus, without the flex-hand approach, the decision maker would possibly not have chosen the selected design.

To conclude, the flex-hand approach takes advantage of the two-stage nature of DESS to automatically select one single design which provides a high flexibility to adapt operation to all considered criteria. Hence, the flex-hand approach importantly contributes to the be-rebust framework of DESS: The flex-hand approach enables the *automatic selection* of the best sustainable design.

Chapter 8

The best reliable and robust sustainable (be-rebust) design

In this chapter, we propose the *be-rebust framework*. To complete the be-rebust framework, we unify the proposed approaches (Fig. 8.1, Section 8.1). The framework enables to select one design which ensures security of energy supply. The selected design is highly flexible regarding the considered aspects of sustainability. For this purpose, uncertain component availability and uncertain energy demands are regarded simultaneously during the multi-objective optimization of DESS. In particular, in a first step, we combine $(n - 1^{max})$-reliability (Section 5.1.2) with RoMO (Section 6.2). The resulting Pareto front obtains reliable and robust options for a design of sustainable DESS. To help the decision maker to select the best design, the flex-hand approach (Section 7) is employed in a second step. For this

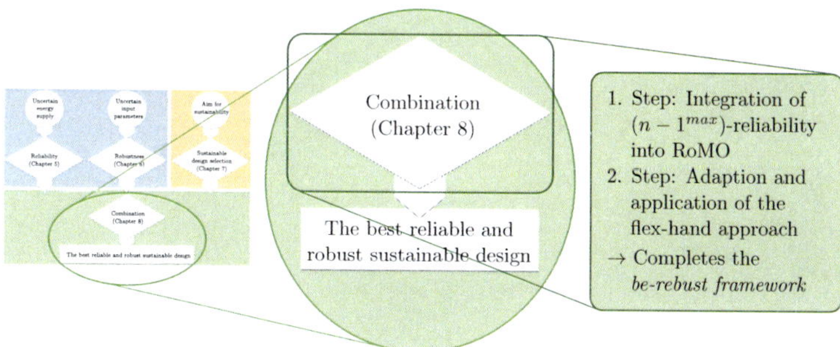

Figure 8.1: Overview of Chapter 8 – the be-rebust framework

purpose, the flex-hand approach needs to be adapted to account for uncertainties (Section 8.1.3). As a result, the selected design is highly flexible regarding the economic and environmental objective functions. The results of the application of the proposed framework are discussed in Section 8.2. A summary and conclusions are provided in Section 8.3.

An alternative combination of the proposed approaches is not recommended to achieve a secure energy supply while regarding aspects of sustainability. Employing $(n-1)$-reliability (Section 5.1.1) or including also the TRusT approach (Section 6.1) would be possible, but drastically increases computational time. In contrast, Chapter 5 shows that $(n-1^{max})$-reliability already identifies well performing reliable designs and enables combination with other approaches. Combining $(n-1^{max})$-reliability only with the TRusT approach without considering the robust global warming impact GWI^{rob} and employing the flex-hand approach would be another alternative; however, taking into account aspects for sustainability, one economic and one environmental objective is chosen in this thesis.

8.1 Reliable robust optimization and the application of the flex-hand approach

In the following, the be-rebust framework is formulated. We expound how we unify the proposed approaches to complete the framework in Section 8.1.1. In a first step of unification, $(n-1^{max})$-reliability is integrated in RoMO (Section 8.1.2). In a second step, to select the best solution, the flex-hand approach is employed (Section 8.1.3).

8.1.1 Unification scheme

We describe the unification scheme consisting of two steps to obtain the framework proposed in this thesis.

1. Step: Integration of $(n-1^{max})$-reliability into RoMO

We start with the schematic problem formulation of RoMO $\left(\mathcal{MP}^{rob}\right)$

$$
\begin{aligned}
\left(\mathcal{MP}^{rob}\right) \quad \min \quad & \begin{pmatrix} \alpha^{TAC} \\ \alpha^{GWI} \end{pmatrix} \\
\text{s.t.} \quad & \text{objective functions} \\
& TAC(x^f, x^s, \xi) \leq \alpha^{TAC} && \forall \xi \in \mathcal{U} \\
& GWI(x^s, \xi) \leq \alpha^{GWI} && \forall \xi \in \mathcal{U} \\
& \text{energy balances and further constraints} \\
& x^f \in \mathcal{X}^f(\xi) && \forall \xi \in \mathcal{U} \\
& x^s \in \mathcal{X}^s(x^f, \xi) && \forall \xi \in \mathcal{U},
\end{aligned}
$$

where $\mathcal{X}^f(\xi)$ and $\mathcal{X}^s(x^f, \xi)$ are the feasible sets of first-stage variables x^f and second-stage variables x^s which depend on the scenario ξ of the uncertainty set $\mathcal{U}$. To introduce reliability into RoMO, additional constraints related to the energy balances need to be regarded (Section 5.1). We denote these additional constraints by

$$F^{rely}(x^f, x^s) \geq 0\,.$$

As we claim for reliability and robustness at the same time, the additional constraints for reliability must still be valid for any uncertain scenario $\xi \in \mathcal{U}$. Hence, the following constraints must be valid to ensure reliability *and* robustness:

$$F^{rely}(x^f, x^s, \xi) \geq 0 \qquad \forall \xi \in \mathcal{U}\,. \tag{8.1}$$

We introduce the additional constraints (Eq. (8.1)) into the RoMO problem formulation $\left(\mathcal{MP}^{rob}\right)$ and obtain the reliable and robust problem formulation $(\mathcal{MP}^{rr})$. The superscript rr emphasizes that the design options are reliable and robust:

$$
\begin{aligned}
(\mathcal{MP}^{rr}) \quad \min \quad & \begin{pmatrix} \alpha^{TAC} \\ \alpha^{GWI} \end{pmatrix} \\
\text{s.t.} \quad & \text{objective functions} \\
& TAC(x^f, x^s, \xi) \leq \alpha^{TAC} && \forall \xi \in \mathcal{U} \\
& GWI(x^s, \xi) \leq \alpha^{GWI} && \forall \xi \in \mathcal{U} \\
& \text{reliability constraints with integrated robustness} \\
& F^{rely}(x^f, x^s, \xi) \geq 0 && \forall \xi \in \mathcal{U} \\
& \text{energy balances and further constraints} \\
& x^f \in \mathcal{X}^f(\xi) && \forall \xi \in \mathcal{U} \\
& x^s \in \mathcal{X}^s(x^f, \xi) && \forall \xi \in \mathcal{U}.
\end{aligned}
$$

By solving the problem ($\mathcal{MP}^{rr}$) we obtain a Pareto front with changing design options which are reliable and robust. We call the resulting Pareto front *ideal reliable and robust Pareto front* (in short: *ideal*^{rr} *Pareto front*). The respective design options are called *ideal*^{rr} *designs*. To select one single design, we employ the flex-hand approach in a second step.

2. Step: Adaption and application of the flex-hand approach

To select one single design which ensures security of energy supply, reliability and robustness need to be integrated into the flex-hand approach. First, we consider the schematic problem formulation of the flex-hand problem ($\mathcal{FP}$):

$$
\begin{aligned}
(\mathcal{FP}) \quad \min \quad & \bar{\varepsilon} \\
\text{s.t.} \quad & \text{objective functions} \\
& \overline{TAC}(x^f, x^s, j) - \overline{TAC}^*(j) \leq \bar{\varepsilon} && \forall j \in [N] \\
& \overline{GWI}(x^s, j) - \overline{GWI}^*(j) \leq \bar{\varepsilon} && \forall j \in [N] \\
& \text{energy balances and further constraints} \\
& x^f \in \mathcal{X}^f \\
& x^s(j) \in \mathcal{X}^s(x^f, j) && \forall j \in [N].
\end{aligned}
$$

The total annualized costs $TAC^*(j)$ and the global warming impact $GWI^*(j)$ refer to point $j \in [N]$ on the ideal Pareto front. Again, bars denote normalized values.

As we aim for reliability and robustness, additional constraints need to be added as performed in the first step (Eq. 8.1). Since the operation needs to be adaptable for each point $j \in [N]$ on the Pareto front, we introduce a dependency of point j:

$$
F^{rely}(x^f, x^s, \xi, j) \geq 0 \qquad \forall \xi \in \mathcal{U}, j \in [N]. \tag{8.2}
$$

To achieve robustness, we need to consider robust objective functions. Hence, the supremum of the objective values is determined by the auxiliary variable α. Furthermore, the selection of one design is based on the ideal reliable and robust Pareto front computed by solving problem ($\mathcal{MP}^{rr}$) in the fist step. Hence, to measure the distance, we employ the normalized objective function values

$\left(\overline{TAC^{rr}}\right)^*(j)$ and $\left(\overline{GWI^{rr}}\right)^*(j)$ instead of $\overline{TAC}^*(j)$ and $\overline{GWI}^*(j)$, respectively:

$$\begin{aligned} \min \quad & \bar{\varepsilon} \\ \text{s.t.} \quad & \text{objective functions} \\ & \overline{\alpha^{TAC}}(j) - \left(\overline{TAC^{rr}}\right)^*(j) \leq \bar{\varepsilon} && \forall j \in [N] \\ & \overline{\alpha^{GWI}}(j) - \left(\overline{GWI^{rr}}\right)^*(j) \leq \bar{\varepsilon} && \forall j \in [N] \\ & TAC(x^f, x^s, \xi, j) \leq \alpha^{TAC}(j) && \forall \xi \in \mathcal{U}, j \in [N] \\ & GWI(x^s, \xi, j) \leq \alpha^{GWI}(j) && \forall \xi \in \mathcal{U}, j \in [N]. \end{aligned} \tag{8.3}$$

As all constraints need to be fulfilled for all scenarios, the feasible sets $\mathcal{X}^f$ and $\mathcal{X}^s$ also depend on scenario $\xi \in \mathcal{U}$. We include the additional constraints for reliability (Eq. (8.2)) and substitute the objective function as well as the feasible sets. Thereby, we obtain the problem formulation $(\mathcal{FP}^{rr})$ to select one reliable and robust design:

$$\begin{aligned} (\mathcal{FP}^{rr}) \quad \min \quad & \bar{\varepsilon} \\ \text{s.t.} \quad & \text{objective functions} \\ & \overline{\alpha^{TAC}}(j) - \left(\overline{TAC^{rr}}\right)^*(j) \leq \bar{\varepsilon} && \forall j \in [N] \\ & \overline{\alpha^{GWI}}(j) - \left(\overline{GWI^{rr}}\right)^*(j) \leq \bar{\varepsilon} && \forall j \in [N] \\ & TAC(x^f, x^s, \xi, j) \leq \alpha^{TAC}(j) && \forall \xi \in \mathcal{U}, j \in [N] \\ & GWI(x^s, \xi, j) \leq \alpha^{GWI}(j) && \forall \xi \in \mathcal{U}, j \in [N] \\ & \text{reliability constraints with integrated robustness} \\ & F^{rely}(x^f, x^s, \xi, j) \geq 0 && \forall \xi \in \mathcal{U}, j \in [N] \\ & \text{energy balances and further constraints} \\ & x^f \in \mathcal{X}^f(\xi) && \forall \xi \in \mathcal{U} \\ & x^s(j) \in \mathcal{X}^s(x^f, \xi, j) && \forall \xi \in \mathcal{U}, j \in [N]. \end{aligned}$$

8.1.2 Integration of $(n-1^{max})$-reliability into RoMO

To integrate $(n-1^{max})$-reliability (Section 5.1.2) into RoMO (Section 6.2), only additional constraints need to be added to the problem formulation of RoMO as explained in Section 8.1.1. The additional $(n-1^{max})$-reliability constraints are based on Eq.s (5.8),(5.10)-(5.12) of Chapter 5. The equations are restated in the following for improved comprehensibility:

$$\sum_{k \in \mathcal{K}^{h/c}} \dot{V}_k^N - \dot{V}_{\mathcal{K}^{h/c}}^{max} \geq \dot{E}_t^{h,tot/c} \qquad \forall t \in \mathcal{T} \tag{5.8}$$

$$\dot{V}_{\mathcal{K}^{h/c}}^{max} \geq \dot{V}_k^N \qquad \forall k \in \mathcal{K}^{h/c} \tag{5.10}$$

$$\sum_{k\in\mathcal{K}^h} \dot{V}_k^N \geq \dot{E}_t^h + \sum_{k\in\mathcal{AC}} \frac{\dot{V}_{kt}}{\eta_k} + \dot{E}_{\mathcal{AC}}^h \qquad \forall t \in \mathcal{T} \tag{5.11}$$

$$\dot{E}_{\mathcal{AC}}^h = \min\left\{ \max_{k\in\mathcal{CC}} \frac{\dot{V}_k^N}{\eta_{\mathcal{AC}}^{min}} \; ; \; \sum_{k\in\mathcal{AC}} \frac{\dot{V}_k^N}{\eta_k} \right\}. \tag{5.12}$$

Following the unification scheme (Section 8.1.1), we integrate robustness into the reliability constraints to ensure security of energy supply during failure of components and uncertain demands.

Hence, the dependency on the uncertain parameters needs to be incorporated as demonstrated in Eq. (8.1). Eq.s (5.10) and (5.12) do not directly depend on the uncertain demands and thus can be integrated without any changes. The MILP reformulation of maximal additionally needed heating supply due to failure of a cooling component $\dot{E}_{\mathcal{AC}}^h$ remains identical compared to Chapter 5 and is provided in Appendix B.2.1. As a result, we only need to adapt Eq.s (5.8) and (5.11), as performed in the following:

$$\sum_{k\in\mathcal{K}^{h/c}} \dot{V}_k^N - \dot{V}_{\mathcal{K}^{h/c}}^{max} \geq \tilde{\dot{E}}_t^{h,tot/c} \qquad \forall t \in \mathcal{T}, \xi \in \mathcal{U} \tag{8.4}$$

$$\sum_{k\in\mathcal{K}^h} \dot{V}_k^N \geq \tilde{\dot{E}}_t^h + \sum_{k\in\mathcal{AC}} \frac{\dot{V}_{kt}}{\eta_k} + \dot{E}_{\mathcal{AC}}^h \qquad \forall t \in \mathcal{T}, \xi \in \mathcal{U} \tag{8.5}$$

which can be replaced by

$$\sum_{k\in\mathcal{K}^c} \dot{V}_k^N - \dot{V}_{\mathcal{K}^c}^{max} \geq \hat{\dot{E}}_t^c + \delta_t^{\dot{E}c} \qquad \forall t \in \mathcal{T} \tag{8.6}$$

$$\sum_{k\in\mathcal{K}^h} \dot{V}_k^N - \dot{V}_{\mathcal{K}^h}^{max} \geq \hat{\dot{E}}_t^h + \delta_t^{\dot{E}h} + \sum_{k\in\mathcal{AC}} \frac{\dot{V}_{kt}}{\eta_k} \qquad \forall t \in \mathcal{T} \tag{8.7}$$

$$\sum_{k\in\mathcal{K}^h} \dot{V}_k^N \geq \hat{\dot{E}}_t^h + \delta_t^{\dot{E}h} + \sum_{k\in\mathcal{AC}} \frac{\dot{V}_{kt}}{\eta_k} + \dot{E}_{\mathcal{AC}}^h \qquad \forall t \in \mathcal{T}. \tag{8.8}$$

As a result, the DESS can supply sufficient energy during a failure of one arbitrary component even if the demands are at their upper bounds.

As explained in Section 6.1.4, overproduction might be necessary to be able to cover the heating demands: There might exist a scenario ξ' for which the *total* heating demand $\dot{E}_t^{h'} + \sum_{k\in\mathcal{AC}} \frac{\dot{V}'_{kt}}{\eta_k}$ is higher than the total heating demand for the upper bound $\hat{\dot{E}}_t^h + \delta_t^{\dot{E}h} + \sum_{k\in\mathcal{AC}} \frac{\dot{V}_{kt}}{\eta_k}$ in the respective optimal operation modes. This case is rather unlikely in practice but may occur if the heating demand of the absorption chillers $\sum_{k\in\mathcal{AC}}\frac{\dot{V}'_{kt}}{\eta_k}$ is significantly higher than in the optimal operation mode for the upper bound $\sum_{k\in\mathcal{AC}}\frac{\dot{V}_{kt}}{\eta_k}$. Even though the scenario ξ' is not directly considered in the optimization, the demands can still be fulfilled. To cover the demands of scenario ξ', the system could be operated in the operation mode for the upper bound of the demands $\widehat{\dot{E}}_t^h + \delta^{\dot{E}h}$ while overproducing heat. In this operation mode the heating demand of the absorption chillers $\sum_{k\in\mathcal{AC}}\frac{\dot{V}_{kt}}{\eta_k}$ is reduced compared to the optimal operation mode in scenario ξ'. However, a sufficient cooling supply is still ensured by Eq. (8.6).

To reduce overproduction, we additionally consider the nominal case and the lower bound as in the TRusT approach and as in RoMO (Chapter 6). For enhanced comprehensibility in the unification scheme (Section 8.1.1), we did not consider the nominal case and the lower bound in the schematic explanations. However, the adaption of constraints follows the same scheme as for Eq.s (8.7) and (8.8):

$$\sum_{k\in\mathcal{K}^h}\dot{V}_k^N - \dot{V}_{\mathcal{K}^h}^{max} \geq \widehat{\dot{E}}_t^h + \sum_{k\in\mathcal{AC}}\frac{\widehat{\dot{V}}_{kt}}{\eta_k} \qquad \forall t\in\mathcal{T} \tag{8.9}$$

$$\sum_{k\in\mathcal{K}^h}\dot{V}_k^N \geq \widehat{\dot{E}}_t^h + \sum_{k\in\mathcal{AC}}\frac{\widehat{\dot{V}}_{kt}}{\eta_k} + \dot{E}_{\mathcal{AC}}^h \qquad \forall t\in\mathcal{T} \tag{8.10}$$

$$\sum_{k\in\mathcal{K}^h}\dot{V}_k^N - \dot{V}_{\mathcal{K}^h}^{max} \geq \widehat{\dot{E}}_t^h - \delta_t^{\dot{E}h} + \sum_{k\in\mathcal{AC}}\frac{\dot{V}_{kt}}{\eta_k} \qquad \forall t\in\mathcal{T} \tag{8.11}$$

$$\sum_{k\in\mathcal{K}^h}\dot{V}_k^N \geq \widehat{\dot{E}}_t^h - \delta_t^{\dot{E}h} + \sum_{k\in\mathcal{AC}}\frac{\dot{V}_{kt}}{\eta_k} + \dot{E}_{\mathcal{AC}}^h \qquad \forall t\in\mathcal{T}\,. \tag{8.12}$$

Since the constraint related to reliability of the cooling circuit (Eq. (8.6)) does not contain any operational variables, the respective nominal case and the lower bound do not need to be considered.

Adding constraints of Eq.s (5.8),(5.11),(8.6)-(8.12) to the problem formulation of RoMO, unifies reliability and robustness. The full problem formulation to obtain reliable and robust design options of DESS is proposed in Appendix E.1.1.

8.1.3 Adaption and application of the flex-hand approach

To employ the flex-hand approach (Chapter 7) to select the be-rebust design, reliability as well as robustness need to be included in the problem formulation in a second step, as explained in Section 8.1.1. Reliability and robustness regarding the energy demands can be included similar to Section 8.1.2. However, the robust second-stage variables need to be able to adapt operation for each point j on the flex-hand Pareto front in order to minimize the distance to the ideal^{rr} Pareto front. Thus, the variables depend on points $j \in [N]$ of the Pareto front as demonstrated in Eq. (8.2). The reliability constraints combined with robustness of Eq.s (5.10),(5.12),(8.6) do not depend on the second-stage variables and, thus, are not adapted. Furthermore, the nominal operation and the operation for the lower bound of the energy demands do not directly affect the objective functions and thus the values of the flex-hand Pareto front. As a result, an adaption to each point on the ideal^{rr} Pareto front does not need to be considered; thus, Eq.s (8.9)-(8.12) are not adapted. Only Eq.s (8.7) and (8.8) need to be reformulated as these constraints directly influence the flex-hand Pareto front by the operational variables $\dot{V}_{kt}(j)$:

$$\sum_{k \in \mathcal{K}^h} \dot{V}_k^N - \dot{V}_{\mathcal{K}^h}^{max} \geq \hat{\dot{E}}_t^h + \delta_t^{\dot{E}h} + \sum_{k \in \mathcal{AC}} \frac{\dot{V}_{kt}(j)}{\eta_k} \qquad \forall t \in \mathcal{T}, j \in [N]$$

$$\sum_{k \in \mathcal{K}^h} \dot{V}_k^N \geq \hat{\dot{E}}_t^h + \delta_t^{\dot{E}h} + \sum_{k \in \mathcal{AC}} \frac{\dot{V}_{kt}(j)}{\eta_k} + \dot{E}_{\mathcal{AC}}^h \qquad \forall t \in \mathcal{T}, j \in [N].$$

Besides the additional equations for reliability and robustness regarding the energy demands, robustness needs to be integrated in the objective functions as performed in Eq. (8.3):

$$\begin{aligned}
\min \quad & \bar{\varepsilon} \\
\text{s.t.} \quad & \text{objective functions} \\
& \overline{\alpha^{TAC}}(j) - \left(\overline{TAC^{rr}}\right)^*(j) \leq \bar{\varepsilon} && \forall j \in [N] \\
& \overline{\alpha^{GWI}}(j) - \left(\overline{GWI^{rr}}\right)^*(j) \leq \bar{\varepsilon} && \forall j \in [N] \\
& \sum_{t \in \mathcal{T}} \Big[\Delta\tau_t \Big(\widehat{p}^{gas}(1+\delta^{pg}) \cdot \sum_{k \in \mathcal{B} \cup \mathcal{CHP}} \dot{U}_{kt}(j) \\
& \qquad + \widehat{p}^{el,buy}(1+pe) \cdot \dot{U}_t^{el,buy}(j) \\
& \qquad - \widehat{p}^{el,sell}(1+pe) \cdot \dot{V}_t^{el,sell}(j) \Big) \Big] \\
& \qquad + \sum_{k \in \mathcal{K}} \left(\frac{1}{PVF} + p_k^m \right) \cdot I_k \leq \alpha^{TAC}(j) && \forall pe \in \{-\delta^{pe}, \delta^{pe}\} \\
& && \forall j \in [N]
\end{aligned}$$

$$\sum_{t \in \mathcal{T}} \Delta\tau_t \Bigg[\sum_{k \in \mathcal{B} \cup \mathcal{CHP}} \dot{U}_{kt}(j) \cdot GWI^{gas}$$

$$+\left(\dot{U}_t^{el,buy}(j) - \dot{V}_t^{el,sell}(j)\right) \cdot \left(\widehat{GWI}^{el} + ge\right)\Bigg] \leq \alpha^{GWI}(j) \quad \begin{array}{r} \forall ge \in \{-\underline{\delta}^{ge}, \overline{\delta}^{ge}\} \\ \forall j \in [N]\,. \end{array}$$

Bars above the total annualized costs TAC and the global warming impact GWI in the optimization problem denote the normalization of the objective values. Objective values on the normalized ideal^{rr} Pareto fronts are denoted by $\left(\left(\overline{TAC}^{rr}\right)^*(j), \left(\overline{GWI}^{rr}\right)^*(j)\right)$ for each point $j \in [N]$. The superscript rr indicates values based on reliability and robustness considerations.

To obtain the full problem formulation, robustness needs to be integrated into further constraints such as limitations of the minimal and maximal installed capacity. The reformulation is given in Appendix E.1.2 where we propose the full problem formulation for the selection of the be-rebust design.

8.2 Results of the case study

The framework stated in the previous Section 8.1 is applied to the real-world case study (Section 4.3) to select the be-rebust design. Again, we assume a "green field" without existing components and the price for purchasing electricity in the nominal scenario is assumed to be $\widehat{p}^{gas} = 6\,\text{ct/kWh}$. To solve the multi-objective optimization problems, we use the adaptive normal boundary intersection method (Das and Dennis, 1998) and employ the solver CPLEX 12.6.3.0 (IBM Corporation, 2015). Calculation to machine accuracy is performed on a computer with 3.24 GHz and 64 GB RAM employing 4 threads. The problem is formulated in GAMS 24.7.3 (McCarl and Rosenthal, 2016). All values regarding model scales and computational times relate to the problem with the largest superstructure which is solved applying the automated superstructure-generation approach from Voll et al. (2013).

8.2.1 The be-rebust design of distributed energy supply systems

Employing the framework enables to select the be-rebust design based on the ideal^{rr} Pareto front (Fig. 8.2). Calculation of the ideal^{rr} Pareto front takes 3 minutes and 12 seconds. Selecting the be-rebust design based on the ideal^{rr} Pareto front, takes 8 minutes and 30 seconds solving a problem with 26 468 constraint, 7 961 variables, and 4 242 binary variables after presolve.

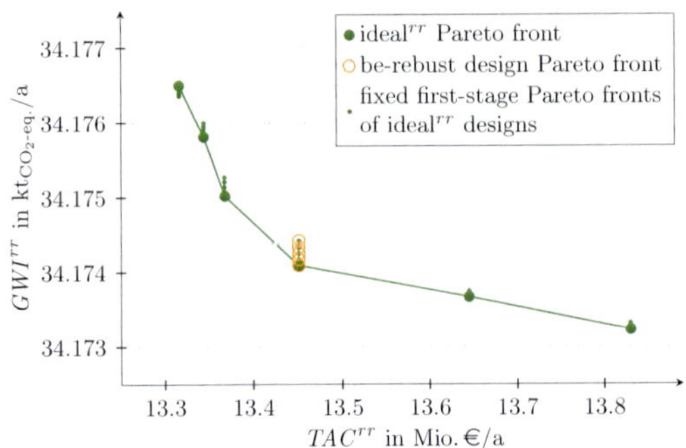

Figure 8.2: Comparison of ideal^{rr} Pareto front (dark green dots) and the Pareto front based on the be-rebust design (be-rebust design Pareto front; orange circles); small dark green dots: fixed first-stage Pareto front of ideal^{rr} designs, i. e., efficient operation for each design of the ideal^{rr} Pareto front; rr highlights that solutions are reliable and robust; Pareto fronts are presented without normalization; lines are included to guide reader's eyes

The fixed first-stage Pareto fronts as well as the be-rebust design Pareto front have only marginally changes in the total annualized costs TAC^{rr} along their Pareto fronts. The be-rebust design leads to similar values as one design option on the ideal^{rr} Pareto front, even though the designs differ. The be-rebust design and the designs on the ideal^{rr} Pareto front are presented in Appendix E.2.1. The normalized minimal distance $(\overline{\varepsilon}^{rr})^*$ is 0.26. Compared to the flex-hand design without regarding reliability and robustness (Section 7.3), the value for the normalized minimal distance $(\overline{\varepsilon}^{rr})^*$ is significantly larger. A larger value implies a reduced flexibility of the be-rebust design to adapt operation to the ideal^{rr} Pareto front.

Having a closer look at the global warming impact GWI^{rr} reveals that there is no significant trade-off between the criteria, since the variation on the y-axis is only 0.03 %. The reason for the small variation of the global warming impact GWI^{rr} is the same as for the reduced trade-off in the Pareto front of RoMO (Section 6.2.3): The consideration of the upper and lower bound of the uncertain specific global warming impact $\widetilde{GWI}^{el}$ of the electricity mix leads to a system striving autonomy from the electricity grid. Since there is no significant trade-off, solutions with smaller total annualized costs TAC^{rr} than costs for the be-rebust design might be more relevant than the selected design. Hence, an application of the adapted flex-hand approach of the be-rebust framework is only recommended, if there is a clear trade-of between the regarded criteria. In general, we expect to have larger trade-offs for energy systems employing renewable energies (Section 6.2.3; Mayer (2018)). However, we propose an idea of an alternative normalization which allows for small variations of values for one criterion in the flex-hand approach.

Alternative normalization of Pareto fronts in the flex-hand approach

Instead of employing the normalization proposed in Eq. (7.2). The objective functions can be normalized only by their maximal values:

$$\overline{f}_i(x^f, x^s) = \frac{f_i(x^f, x^s)}{\max_{j \in [N]} \rho_i^*(j)} .$$

Employing this normalization, does not distort the shape of the Pareto front and correlates the variation of values for the criteria to their influence on the selected design. Hence, the selection of the design is less dependent of criteria comprising a small variation of values.

A first application shows the desired effect: The influence of the criteria with small variation, i. e., the global warming impact GWI^{rr}, is reduced and the selected design generates objective values close to the minimal total annualized costs TAC^{rr} (Fig. 8.3).

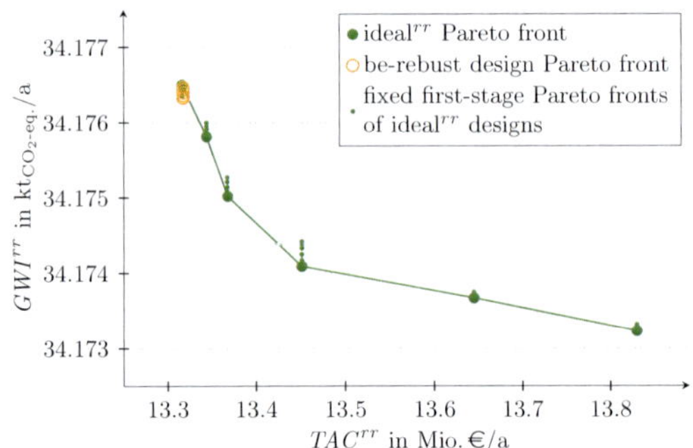

Figure 8.3: Comparison of ideal^{rr} Pareto front (dark green dots) and the Pareto front based on the be-rebust design (orange circles) for alternative normalization; small dark green dots: fixed first-stage Pareto front of ideal^{rr} designs, i. e., Pareto-efficient operation for each design of the ideal^{rr} Pareto front; rr highlights that solutions are reliable and robust; Pareto fronts are presented without normalization; lines are included to guide reader's eyes

8.2.2 Sensitivity analysis of the be-rebust design

As upper and lower bounds of the uncertainties are inherent uncertain themselves (Section 6.1.5) a sensitivity analysis on the uncertainty of input parameters is performed (Fig.8.4[1]). Again, we use ω to scale all uncertainties ($\omega \in \{0.33, 0.66, 1, 1.33, 1.66\}$).

For uncertainty-scaling factors $\omega \geq 1$, no significant trade-off can be observed. For scaling factor $\omega \leq 0.66$, a trade-off occurs. As discussed in Section 6.2.3, regarding the upper and lower bound of the uncertain specific global warming impact of the electricity grid mix $\widetilde{GWI}^{el}$ leads to a system which is autonomous from the electricity grid causing a nearly constant global warming impact GWI^{rr}. In contrast, when decreasing the uncertainty of the specific global warming impact

[1]The automated superstructure-generation approach from Voll et al. (2013) is terminated for the calculation with $\omega = 0.33$ after 8 components were regarded, since solving the problem regarding 9 components was not possible within 72 h. For comparison, solving the problem regarding 8 components took only 3 minutes and 40 seconds. As a result, the flex-hand Pareto front is not solved to optimality and hence no operation strategy can totally reach the ideal Pareto front. However, the calculated design already performs well providing high flexibility.

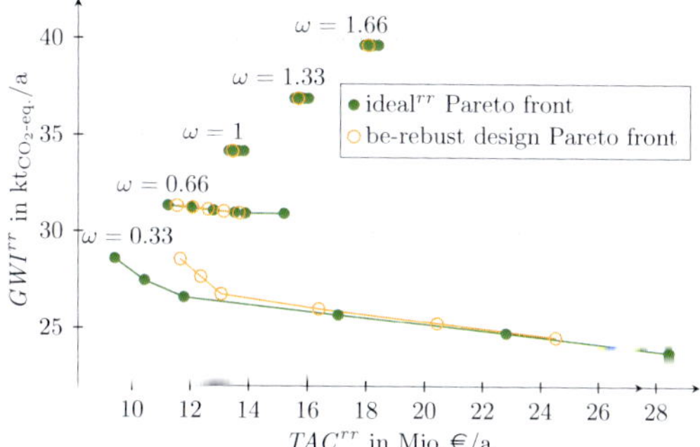

Figure 8.4: Sensitivity analysis on input parameter uncertainty with respect to the uncertainty-scaling factor ω; comparison of ideal^{rr} Pareto front (dark green dots) and the be-rebust design Pareto front (orange circles); rr highlights that solutions are reliable and robust; Pareto fronts are presented without normalization; lines are included to guide reader's eyes

$\widetilde{GWI}^{el}$ (employing factors $\omega = 0.33$ and $\omega = 0.66$), the trade-off increases. As the lower bound of the specific global warming impact $\widetilde{GWI}^{el}$ increases, selling electricity to the grid gets increasingly beneficial for the global warming impact GWI^{rr}. For uncertainty-scaling factors $\omega = 0.33$ and $\omega = 0.66$, the employed framework enables to select highly flexible designs.

Remarkable is the performance of the be-rebust design for $\omega = 0.66$ (Fig. 8.5). The normalized minimal distance $(\bar{\varepsilon}^{rr})^*$ is only 0.08 which represents the high adaptability of the be-rebust design to the ideal^{rr} Pareto front. Selecting designs with small global warming impact GWI^{rr} leads to a drastic increase of the global warming impact GWI^{rr} if aims change and small total annualized costs TAC^{rr} are desired. E. g., for the anchor point with minimal global warming impact GWI^{rr} on the ideal^{rr} Pareto front the global warming impact GWI^{rr} increases from 30.9 $kt_{CO_2\text{-eq.}}/a$ up to 36.0 $kt_{CO_2\text{-eq.}}/a$ which corresponds to an increase of 16.5 %. However, the desired savings in the total annual costs TAC^{rr} are only small with 3.3 %. This effect results from the fact that feeding in electricity supplied by the combined heat and power engines into the grid is beneficial for the global warming impact GWI^{rr} but not for the total annualized costs TAC^{rr}: Trigeneration reduces the specific global warming impact compared to the impact

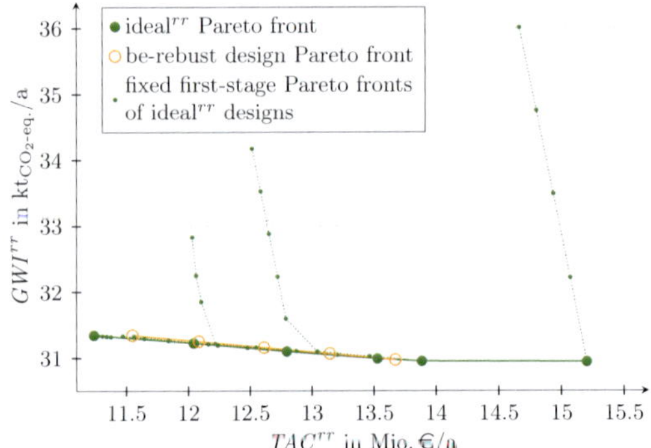

Figure 8.5: Comparison of ideal^{rr} Pareto front (dark green dots) and the Pareto front based on the be-rebust design (orange circles) for uncertainty-scaling factor $\omega = 0.66$; small dark green dots: fixed first-stage Pareto front of ideal^{rr} designs, i. e., efficient operation for each design of the ideal^{rr} Pareto front; rr highlights that solutions are reliable and robust; Pareto fronts are presented without normalization; lines are included to guide reader's eyes

when separately purchasing electricity from the grid and supplying heating energy by boilers; in contrast, the revenues from selling electricity to the grid are not high enough compared to the costs of operating the combined heat and power engines. Thus, with decreasing total annualized costs TAC^{rr} the provided heating energy is supplied by boilers instead of combined heat and power engines while feeding in less electricity into the grid. Two of the fixed first-stage Pareto fronts of the ideal^{rr} designs increase only slightly in the global warming impact GWI^{rr} with decreasing total annualized costs TAC^{rr} at first, following the trend of the ideal^{rr} Pareto front; then, the gradient changes significantly and the global warming impact GWI^{rr} drastically increases (Fig. 8.5). The drastic increase is due to the same effect as for the ideal^{rr} design of the anchor point: Boilers are favored compared to combined heat and power engines. In the part of slightly increasing global warming impact GWI^{rr}, there is a shift between absorption chillers and compression chillers: With increasing global warming impact GWI^{rr}, less absorption chillers are employed and more cooling energy is supplied by the compression chillers. Hence, electricity supplied on site can directly be used and is less fed into the grid. At the point where the gradient changes significantly, the compression chillers are operated at their upper bounds and no further replacement of absorption chillers is possible. To further decrease the total annualized costs TAC^{rr}, the electricity fed into the grid by the combined heat and power engines is significantly reduced and boilers are favored as explained above. In contrast to the ideal^{rr} designs of these two Pareto fronts with at first slight and than drastic increase in the global warming impact GWI^{rr}, the ideal^{rr} design of the anchor point with minimal global warming impact GWI^{rr} does not comprise any absorption chillers on Site A (Fig. 8.6, design on the respective right side); thus, no shifting of cooling supply is possible. As a result, the global warming impact GWI^{rr} drastically increases along the whole Pareto front while replacing combined heat and power engines by boilers in operation. Employing the be-rebust design, boilers supply constant heating energy along the whole Pareto front. Only the shift from absorption chillers to compression chillers occurs inducing only a small increase of the global warming impact GWI^{rr}. The results show that the proposed framework is necessary to select the be-rebust design providing high flexibility.

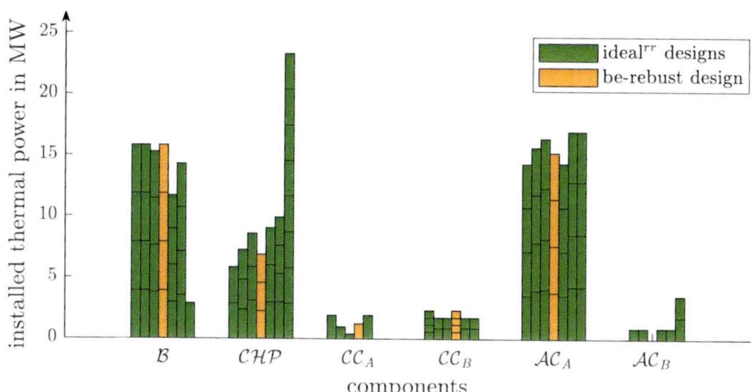

Figure 8.6: In dark green: designs of ideal^{rr} Pareto front; in orange: the be-rebust design with respect to uncertainty-scaling factor $\omega = 0.66$; from left to right for each component type, the designs are ordered by decreasing minimal global warming impact; $\mathcal{B}$ boiler, $\mathcal{CHP}$ combined heat and power engine, $\mathcal{CC}_A$ and $\mathcal{CC}_B$ compression chillers, and $\mathcal{AC}_A$ and $\mathcal{AC}_B$ absorption chillers installed on Site A and Site B, respectively; rr highlights that solutions are reliable and robust

The installed thermal power of combined heat and power engines $\mathcal{CHP}$ increases with decreasing global warming impact GWI^{rr}, while the installed thermal power of boilers $\mathcal{B}$ decreases (Fig. 8.6). Similar to the results of Section 7.3, no remarkable characteristics of the be-rebust design can be observed. For a presentation of the be-rebust design and the designs on the ideal^{rr} Pareto front for uncertainty-scaling factor $\omega = 0.33$, see Appendix E.2.2.

The framework regards the trade-off between total annualized costs TAC^{rr} and global warming impact GWI^{rr} without considering nominal costs as proposed by the TRusT approach (Section 6.1). To evaluate the performance of the be-rebust design in nominal operation, we re-optimize operation of the be-rebust design without taking into account uncertainties of supply and input parameters (Fig. 8.7). For the performance in the nominal case of the be-rebust design with uncertainty-scaling parameter $\omega = 1$, see Appendix E.2.1.

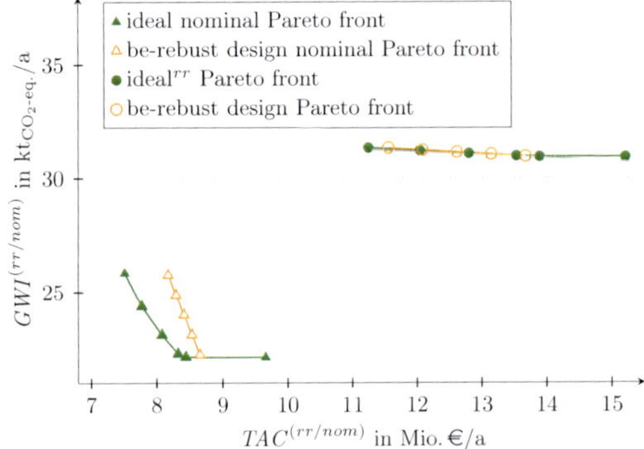

Figure 8.7: Comparison of nominal Pareto front (dark green triangles) and the be-rebust design nominal Pareto front, i. e., the nominal Pareto front based on the be-rebust design, (orange unfilled triangles) with respect to uncertainty-scaling factor $\omega = 0.66$; the ideal^{rr} Pareto front and the be-rebust design Pareto front are presented only for comparison, (see also Fig. E.1); rr highlights that solutions are reliable and robust; Pareto fronts are presented without normalization; lines are included to guide reader's eyes

For uncertainty-scaling parameter $\omega = 0.66$, the comparison of the Pareto fronts shows that the be-rebust design performs well in the nominal operation. Since the be-rebust design includes redundant components to ensure reliability, a certain distance to the ideal nominal total annualized costs TAC^{*} is expected due to increased investment costs. The minimal nominal total annualized costs $TAC^{rr/nom}$ (obtained by nominal operational optimization of the be-rebust design) differs by 8.8 % from the ideal nominal total annualized costs TAC^{*}; the global warming impact differs by only 0.5 %. A combination with the TRusT approach (as in Section 6.3) could possibly further decrease this already low deviation. However, the combination is not performed in this thesis, since computational effort is expected to be high as many additional operational variables and constraints are necessary to adapt operation to every point j on the Pareto front during design selection.

8.3 Conclusions

We answer the question of how to select one single sustainable energy system design which ensures security of energy supply despite uncertain component availability and uncertain energy demands. For this purpose, we combine approaches for reliability, robustness, and sustainable design selection and complete the framework proposed in this thesis. The combination of the approaches follows a consistent unification scheme (Section 8.1.1). In a first step, we unify reliability with robustness while optimizing the total annualized costs and the global warming impact (Section 8.1.2). In particular, we integrate $(n-1^{max})$-reliability (Section 5.1.2) into RoMO (Section 6.2). For integration, only additional constraints for reliability need to be adapted regarding robustness and added to the problem formulation of RoMO. By solving the integrated problem, we obtain a Pareto front with design options which are reliable and robust. Based on this Pareto front, we select one single design in a second step. For this purpose, the flex-hand approach (Chapter 7) is applied. We adapt the flex-hand approach to account for uncertainties of supply and input parameters also during selection of the best design (Section 8.1.3). The finally resulting framework is easy to implement as only constraints and operational variables need to be added following the proposed unification scheme.

The results of the case study show that the proposed framework is an excellent tool to support the decision maker by automatic selection of the be-rebust design. However, if there is no remarkable trade-off between the objective functions, since the variation for one criterion of the Pareto front is only marginal, the application of the flex-hand approach to select one design is not appropriate: The selected design leads to objective values located close to the center of the Pareto front. In case of small variation for one criterion, the other criterion should have a higher influence on decision making which would lead to values shifted to the related anchor point. To eliminate this shortcoming, we propose an idea of how to allow for small variation of values for one criterion in the flex-hand approach (Section 8.2.1).

The performed sensitivity analysis on uncertainty of input parameters reveals a clear trade-off between the objective criteria for reduced uncertainty. Based on the respective Pareto front, the be-rebust framework performs excellent and selects a highly flexible design. In contrast, when choosing an arbitrary design of the given design options of the ideal Pareto front without employing the framework, the corresponding objective value might drastically increase if the prioritization of criteria changes: For instance, implementing the design with minimal global warming impact increases in the global warming impact by 16.5 %, if the chosen design is operated such that minimal total annualized costs are achieved (Section 8.2.2). Hence, the be-rebust framework is necessary to select a design which performs well in operation regarding all considered criteria.

To evaluate the be-rebust design in nominal operation, the operation for the be-rebust design is re-optimized for the nominal problem without regarding any uncertainties of supply and input parameters. The results show that the be-rebust design performs well with a deviation of only 0.5 % between the minimal robust and the minimal nominal global warming impact, and a deviation of 8.8 % between the minimal robust and the minimal nominal total annualized costs. However, the optimal nominal design is neither reliable nor robust; hence, the deviation in the total annualized costs is mainly caused by additional costs for a secure energy supply. To select a design which performs even better in the nominal case, a combination with the TRusT approach as proposed in Section 6.3 would also be possible. However, high computational effort is expected.

As results show, the proposed framework successfully combines the approaches for robustness, reliability, and sustainable design selection. As a result, the framework allows to select a highly flexible sustainable design which is reliable and robust.

Chapter 9

Summary, conclusions, and future perspectives

This chapter summarizes and concludes the findings of this thesis and provides future perspectives of research.

9.1 Summary and conclusions

The synthesis of sustainable energy systems is a challenging task involving multiple criteria. Thus, multi-objective optimization is a highly suitable tool to identify sustainable design options. However, multi-objective optimization leaves the decision maker with multiple options to select from. Hence, the question arises: How to select one design?

The complexity of decision making is further increased as the synthesis problem of energy systems is inherently uncertain. Uncertainty occurs due to uncertain availability of components and uncertain input parameters, such as energy demands and energy prices. Neglecting these uncertainties leads to insecure energy systems, since a lack of energy supply might occur. An insufficient energy supply possibly violates contracts or destroys production units involving high unexpected costs or even environmental damage. Hence, uncertainties of energy supply and input parameters need to be regarded already during design optimization.

This thesis regards both, the selection of one sustainable design *and* the uncertainties in energy system optimization. Hence, this thesis answers the research question stated in Section 3.1:

> *How to select one single sustainable energy system design which ensures security of energy supply despite uncertain component availability and uncertain energy demands?*

For this purpose, the *be-rebust framework* (framework to select the **be**st **re**liable and ro**bust** **s**us**t**ainable design) is proposed in this thesis. The framework combines approaches which regard for reliability, robustness, and sustainable design selection. The proposed approaches takes advantage of the two-stage nature of energy system optimization. The design corresponds to the here-and-and-now decision (= the first stage) and the operation to the wait-and-see decision (= the second stage). Each approach is introduced separately and can also be employed independently:

1. *The reliability approaches*, $(n-1)$-reliability and $(n-1^{max})$-reliability, include uncertainty of energy supply into the framework identifying cost-efficient reliable designs while ensuring energy supply even though 1 arbitrary energy component might fail at any time.
2. *The robustness approaches*, the TRusT (Two-stage Robustness Trade-off) approach and RoMO (minmax Robust Multi-objective Optimization), include uncertainty of input parameters into the framework. At the same time, sustainability is incorporated by regarding economic end environmental decision criteria. Hence, the approaches allow for sufficient energy supply while regarding aspects of sustainability and show that sufficient energy supply is not expensive per se.
3. *The selection approach for sustainable designs*, the flex-hand (flexible here-and-now decision) approach, automatically selects one design based on design options from multi-objective optimization regarding aspects of sustainability. The flex-hand approach enables to select a design which is highly flexible in operation regarding the considered decision criteria.

All approaches can be applied to any two-stage problem where the first stage need to be fixed right now and the second stage can be adapted later. Only $(n-1^{max})$-reliability needs adaption to account for specific characteristics of the optimization problem such as coupling of energy supply circuits.

The proposed approaches are combined to the be-rebust framework. In a first step, reliability is introduced into robust optimization while regarding multiple criteria. For this purpose, we integrate $(n-1^{max})$-reliability into RoMO. Solving this first problem yields a Pareto front containing reliable and robust design options regarding aspects of sustainability. In a second step, based on this Pareto front one design is selected looking even beyond the set of already identified efficient solutions. For this purpose, an adapted formulation of the flex-hand approach is applied taking into account reliability and robustness. The adapted flex-hand approach automatically selects the be-rebust design. The be-rebust design is highly flexible regarding the considered criteria while ensuring security of energy supply.

The developed framework is applied to a real-world case study to select the be-rebust design. To account for sustainability, the total annualized costs and the global warming impact are optimized. The results show that the flex-hand approach selects a highly flexible design for Pareto fronts with a clear trade-off. The high flexibility in operation allows adapting to changing focus within the regarded criteria. As the case study reveals, not all designs on the Pareto front provide such high flexibility in general. Hence, the proposed framework is necessary to select the best sustainable design which ensures security of energy supply.

9.2 Future perspectives

With certainty, this thesis does not finalize research in the area of optimal reliable and robust design of sustainable energy systems. In the following, future research perspectives are given to improve the framework proposed in this thesis and to advance in the research area of optimal synthesis of reliable and robust sustainable energy systems in general. The research ideas are ordered from topics specific for the proposed framework to general research ideas.

Enhancement of the flex-hand approach. As already discussed in Section 8.2.1, the flex-hand approach can be enhanced for Pareto fronts with small variations regarding one criterion. By implementing the proposed idea of adapting the normalization, the influence of the criteria with small variation is reduced. However, a validation of the proposed adaption is necessary. Furthermore, assessing other distance measures might also reveal designs with increased performance.

Deeper understanding of the be-rebust design. As the case study shows, the selected be-rebust design does not show remarkable characteristics compared to the design options on the ideal reliable and robust Pareto front. To achieve a deeper understanding, near-optimal solutions should be analyzed. For this purpose, the SPREAD method by Hennen et al. (2017) could be employed. A deeper understanding of the design would allow for deducing general synthesis rules and is a cornerstone to convince decision makers to accept the computed be-rebust design in practical application.

Definition of secure energy supply. In this thesis, security of energy supply is related to the ability to cover maximal occurring demands during the failure of 1 arbitrary component at any time. Hence, the expression *secure energy supply* is interpreted strictly. Loosening this definition of secure energy supply could improve the ecological criterion. For example, secure supply could be interpreted such that the costs do not increase by more than a certain rang and that no environmental damage occurs. In this context, shifting and shedding of demands could also be taken into account.

Increasing the acceptance. Solutions computed by mathematical optimization often struggle with the acceptance by practical engineers. Besides profoundly expounding the mathematical solution, interdisciplinary research with psychologists is promising to increase acceptance of solutions provided by mathematical optimization.

As emphasized in Chapter 1, "Achieving climate neutrality in the energy sector – while ensuring at the same time a more efficient energy use, a secure supply of energy, affordable prices and low environmental impact – is a complex endeavor" (European Commission, 2018). Hence, the optimal design of reliable and robust sustainable energy systems offers a large unexplored area for research. However, theoretical concepts need to be implemented in practical application. For this, politics are urged to act now to accelerate the transition of energy systems based on current research towards a climate neutral and secure energy sector – since:

> *True optimization is the revolutionary contribution of modern research to decision processes.*
>
> George Bernhardt Dantzig (1904 - 2005)

Appendices

Appendix A

Optimal design

This appendix provides supplemental material for the optimization of DESS.

A.1 Linearized investment costs

The investment costs I_k of a newly installed component k are linearized by piecewise linearization with $\sum_{h\in[H]}\left[\gamma_{kh} \cdot \kappa_{kh} + m_{kh} \cdot \left(\dot{V}_{kh}^{N} - \gamma_{kh}\dot{V}_{kh}^{N,lb}\right)\right]$ (Fig. A.1).

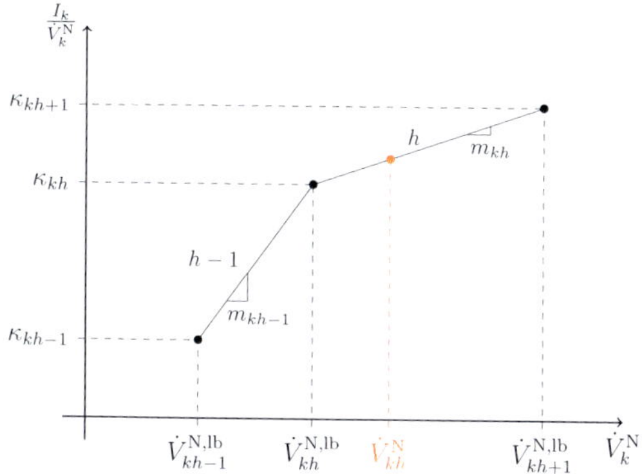

Figure A.1: Piecewise linearization of the investment costs I_k of a newly installed component k; h is the active line segment; thus, the binary variable γ_{kh} is equal to 1

m_{kh} is the slope for each line segment $h \in [H]$ and is defined by

$$m_{kh} := \frac{\kappa_{kh+1} - \kappa_{kh}}{\dot{V}^{N,lb}_{kh+1} - \dot{V}^{N,lb}_{kh}} \quad \forall k \in \mathcal{K}, h \in [H].$$

The parameters $\dot{V}^{N,lb}_{kh+1}$ and $\dot{V}^{N,lb}_{kh}$ represent the nominal capacities of the lower and upper supporting point of line segment h. The parameters κ_{kh} and κ_{kh+1} the corresponding specific investment costs. Binary variables γ_{kh} determine if line segment h is active ($\gamma_{kh} = 1$). The installed thermal power $\dot{V}^{N}_{kh}$ of the components $k \in \mathcal{K}$ for each line segment h is bound by the nominal capacities of the adjoining supporting points, i. e., $\dot{V}^{N,lb}_{kh}$ and $\dot{V}^{N,lb}_{kh+1}$, respectively:

$$\gamma_{kh}\dot{V}^{N,lb}_{kh} \leq \dot{V}^{N}_{kh} \leq \gamma_{kh}\dot{V}^{N,lb}_{kh+1} \qquad h \in \{1, \ldots, n\}.$$

Since we claim $\sum_{h \subset [H]} \gamma_{kh} \leq 1$, only one line segment can be active at the time. Thus, only one value for the installed thermal power $\dot{V}^{N}_{kh}$ of all line segments is unequal to 0; hence, the installed thermal power $\dot{V}^{N}_{k}$ of an installed component k is given by the sum $\sum_{h \in [H]} \dot{V}^{N}_{kh}$ (for the full problem formulation see Appendix A.2).

A.2 Full problem formulation

In the following, we provide the problem formulation of the nominal optimization for the DESS (Section 4.1). We consider the total annualized costs TAC and the global warming impact GWI as objective functions.

$$\min \begin{pmatrix} TAC\left(\dot{U}, \dot{U}^{el,buy}, \dot{V}^{el,sell}, \dot{V}^{N}\right) \\ GWI\left(\dot{U}, \dot{U}^{el,buy}, \dot{V}^{el,sell}\right) \end{pmatrix}$$

s. t. energy balances

$$\sum_{k \in \mathcal{B} \cup \mathcal{CHP}} \dot{V}_{kt} - \sum_{k \in \mathcal{AC}} \dot{U}_{kt} = \dot{E}^{h}_{t} \qquad \forall t \in \mathcal{T}$$

$$\sum_{k \in \mathcal{AC} \cup \mathcal{CC}} \dot{V}_{kt} = \dot{E}^{c}_{t} \qquad \forall t \in \mathcal{T}$$

$$\sum_{k \in \mathcal{CHP}} \dot{V}^{el}_{kt} - \sum_{k \in \mathcal{CC}} \dot{U}_{kt} + \dot{U}^{el,buy}_{t} - \dot{V}^{el,sell}_{t} = \dot{E}^{el}_{t} \qquad \forall t \in \mathcal{T}$$

linearization of investment costs

$$\sum_{h \in [H]} \gamma_{kh} \leq 1 \qquad \forall k \in \mathcal{K}$$

$$\gamma_{kh} \cdot \dot{V}_{kh}^{N,lb} \leq \dot{V}_{kh}^{N} \leq \gamma_{kh} \cdot \dot{V}_{kh+1}^{N,lb} \qquad \forall k \in \mathcal{K}, h \in [H]$$

further constraints and variables

$$\mu^{min} \cdot \sum_{h\in[H]} \dot{V}_{kh}^{N} \leq \dot{V}_{kt} \leq \sum_{h\in[H]} \dot{V}_{kh}^{N} \qquad \forall k \in \mathcal{K}, t \in \mathcal{T}$$

$$\dot{V}_{kt} = \eta_k \cdot \dot{U}_{kt} \qquad \forall k \in \mathcal{K}, t \in \mathcal{T}$$

$$\dot{V}_{kt}^{el} = \eta_k^{tot} \cdot \dot{U}_{kt} - \dot{V}_{kt} \qquad \forall k \in \mathcal{CHP}, t \in \mathcal{T}$$

$$\dot{U}^{el,buy}, \dot{V}^{el,sell}, \dot{V}^{el}, \dot{U}, \dot{V} \in \mathbb{R}_+^{|\mathcal{K}|\times T}$$

$$\gamma \in \{0,1\}^{|\mathcal{K}|\times H}, \dot{V}^N \in \mathbb{R}_+^{|\mathcal{K}|\times H}$$

The objective functions, i. e., the total annualized costs TAC and the global warming impact GWI, are defined by

$$\begin{aligned}
&TAC\left(\dot{U}, \dot{U}^{el,buy}, \dot{V}^{el,sell}, \dot{V}^{N}\right) \\
&= \sum_{t\in\mathcal{T}} \left[\Delta\tau_t \left(p^{gas} \cdot \sum_{k\in\mathcal{B}\cup\mathcal{CHP}} \dot{U}_{kt} + p^{el,buy} \cdot \dot{U}_t^{el,buy} - p^{el,sell} \cdot \dot{V}_t^{el,sell}\right)\right] \\
&\quad + \sum_{k\in\mathcal{K}} \left(\frac{1}{PVF} + p_k^m\right) \cdot \underbrace{\sum_{h\in[H]} \left[\gamma_{kh} \cdot \kappa_{kh} + m_{kh} \cdot \left(\dot{V}_{kh}^{N} - \gamma_{kh}\dot{V}_{kh}^{N,lb}\right)\right]}_{=:I_k} \\
&GWI\left(\dot{U}, \dot{U}^{el,buy}, \dot{V}^{el,sell}\right) \\
&= \sum_{t\in\mathcal{T}} \Delta\tau_t \left[\sum_{k\in\mathcal{B}\cup\mathcal{CHP}} \dot{U}_{kt} \cdot GWI^{gas} + \left(\dot{U}_t^{el,buy} - \dot{V}_t^{el,sell}\right) \cdot GWI^{el}\right].
\end{aligned}$$

The duration of a time step $t \in \mathcal{T}$ is given by $\Delta\tau_t$. Maintenance costs are determined by the share p_k^m of the investment costs I_k. The investment costs I_k are annualized by the present value factor PVF. GWI^{gas} represents the specific global warming impact of purchased gas. Purchased and sold electricity flows are denoted by $\dot{U}_t^{el,buy}$ and $\dot{V}_t^{el,sell}$, respectively. $\dot{U}_{kt}$ and $\dot{V}_{kt}$ specify input and output energy flows in time step t of component $k \in \mathcal{K}$, respectively. The operational variables $\dot{U}_{kt}$, $\dot{V}_{kt}$, $\dot{U}_t^{el,buy}$ and $\dot{V}_t^{el,sell}$ depending on time step t are second-stage variables. The output energy flows need to cover heating demands $\dot{E}_t^h$, cooling demands $\dot{E}_t^c$, and electricity demands $\dot{E}_t^{el}$ for which, however, there is support by the connection to the electricity grid. Components of the trigeneration system include boilers $\mathcal{B}$, combined heat and power engines $\mathcal{CHP}$, absorption chillers $\mathcal{AC}$, and compression chillers $\mathcal{CC}$. The slope m_{kh} for linearization of investment costs is given for all line segments $h \in [H]$ and all components $k \in \mathcal{K}$ by $\frac{\kappa_{kh+1}-\kappa_{kh}}{\dot{V}_{kh+1}^{N,lb}-\dot{V}_{kh}^{N,lb}}$ with given parameters κ_{kh} and $\dot{V}_{kh}^{N,lb}$ (Appendix A.1). Input and output energy flows are coupled by the thermal efficiency η_k. For combined heat and power engines, the total efficiency η_k^{tot} is given by the sum of the thermal and the electrical efficiency $\eta_k^{tot} = \eta_k + \eta_k^{el}$. The minimal part-load of a component k is defined by the fraction μ^{min} of the installed thermal power $\sum_{h\in[H]} \dot{V}_{kh}^{N}$. The installed thermal power $\dot{V}_{kh}^{N}$ and the binary variables γ_{kh} (Appendix A.1) with the corresponding investment costs I_k are first-stage variables.

A.3 Energy demands

Thermal demands of the industrial park also shown in Fig. 4.1 are listed in Tab. A.1 as well as the electricity demands.

Table A.1: Thermal demands (for cooling on Site A and Site B) as well as electricity demands; time step t_1 to t_5 are time-aggregated values; t_6 and t_7 represent minimal and maximal peak demands

demands	t_1	t_2	t_3	t_4	t_5	t_6	t_7
coolA in kW	1446	1780	2540	3137	3982	905	12262
coolB in kW	727	773	844	877	921	467	1714
heat in kW	4272	3308	2444	2333	2289	9463	1473
electricity in kW	5446	5443	5431	5419	5451	7911	7911

Appendix B

Reliability

This appendix provides supplemental material for Chapter 5.

B.1 Motivating example

Let the author take you to your private home for a motivating example (Fig. B.1):

Figure B.1: Motivating example for reliable systems

Imagine you have just bought a perfect lamp for you and your flat – so to say, an optimal lamp. Now, it is getting dark, so you switch the light on. But, suddenly, the light bulb blows – and – it is dark. This incident might lead you to rethink your choice: Maybe your illumination system should not just be optimal but also reliable? Hence, the next day, you install an optimal reliable illumination system comprising 3 small lamps. When it is getting dark, you switch on 2 of the 3 lamps to adequately illuminate your flat. If now 1 light bulb blows, you still got 1 lamp in reserve to achieve sufficient light in your flat.

B.2 The maximal additionally needed heating supply due to failure of a cooling component

In this section, a reformulation of the maximal additionally needed heating supply $\dot{E}^h_{\mathcal{AC}}$ due to failure of a cooling component is proposed to achieve an MILP. Additionally, a tightened formulation is provided.

B.2.1 Reformulation of equation constraining the maximal additional heating $\dot{E}^h_{\mathcal{AC}}$

To obtain an MILP, we reformulate Eq. (5.12) by:

$$\dot{E}^h_{\mathcal{AC}} \geq x \cdot \frac{\dot{V}^{max}_{\mathcal{CC}}}{\eta^{min}_{\mathcal{AC}}} \tag{B.1}$$

$$\eta^{min}_{\mathcal{AC}} \leq \eta_k \qquad \forall k \in \mathcal{AC} \tag{B.2}$$

$$\dot{V}^{max}_{\mathcal{CC}} \geq \dot{V}^N_k \qquad \forall k \in \mathcal{CC} \tag{B.3}$$

$$\dot{E}^h_{\mathcal{AC}} \geq y \cdot \sum_{k \in \mathcal{AC}} \frac{\dot{V}^N_k}{\eta_k} \tag{B.4}$$

$$x + y = 1 \tag{B.5}$$

$$x, y \in \{0, 1\}\,. \tag{B.6}$$

The bilinear products with the binary variables x and y in Eq.s (B.1) and (B.4) can be linearized, e. g., using Golver's linearization (Glover, 1975). The binary variables x and y activate and deactivate the lower bound of the maximal heating supply needed additionally for absorption chillers during the failure of a cooling component $\dot{E}^h_{\mathcal{AC}}$. Since only 1 bound needs to be active (Eq. (B.5)), the maximal additional heating supply $\dot{E}^h_{\mathcal{AC}}$ takes its minimum as claimed in Eq. (5.12).

B.2.2 Tightened formulation of the maximal additional heating $\dot{E}^h_{\mathcal{AC},t}$

The additional heating demand $\dot{E}^h_{\mathcal{AC}}$ (Eq. (5.12)) can be approximated even more accurately, if each time step t is considered separately:

$$\dot{E}^h_{\mathcal{AC},t} = \min \left\{ \max_{k \in \mathcal{CC}} \frac{\dot{V}^N_k}{\eta^{min}_{\mathcal{AC}}} \; ; \; \sum_{k \in \mathcal{AC}} \frac{\dot{V}^N_k}{\eta_k} - \sum_{k \in \mathcal{AC}} \frac{\dot{V}_{kt}}{\eta_k} \; ; \; \frac{\dot{E}^c_t}{\eta^{min}_{\mathcal{AC}}} - \sum_{k \in \mathcal{AC}} \frac{\dot{V}_{kt}}{\eta_k} \right\} . \tag{B.7}$$

The first term remains the same as in Eq. (5.12) (Section 5.1.2). The second term, $\sum_{k \in \mathcal{AC}} \frac{\dot{V}^N_k}{\eta_k} - \sum_{k \in \mathcal{AC}} \frac{\dot{V}_{kt}}{\eta_k}$, reduces the maximal heating energy required by all absorption chillers by the amount of energy which is already reserved for the absorption chillers in the current operation. The third term takes into account the cooling demand on site $\dot{E}^c_t$ which limits the additional heating demand $\dot{E}^h_{\mathcal{AC},t}$ by $\frac{\dot{E}^c_t}{\eta^{min}_{\mathcal{AC}}} - \sum_{k \in \mathcal{AC}} \frac{\dot{V}_{kt}}{\eta_k}$. To obtain an MILP, we reformulate Eq. (B.7) by:

$$\dot{E}^h_{\mathcal{AC},t} \geq x_t \cdot \frac{\dot{V}^{max}_{\mathcal{CC}}}{\eta^{min}_{\mathcal{AC}}} \qquad \forall t \in \mathcal{T} \tag{B.8}$$

$$\eta^{min}_{\mathcal{AC}} \leq \eta_k \qquad \forall k \in \mathcal{AC} \tag{B.9}$$

$$\dot{V}^{max}_{\mathcal{CC}} \geq \dot{V}^N_k \qquad \forall k \in \mathcal{CC} \tag{B.10}$$

$$\dot{E}^h_{\mathcal{AC},t} \geq y_t \cdot \left(\sum_{k \in \mathcal{AC}} \frac{\dot{V}^N_k}{\eta_k} - \sum_{k \in \mathcal{AC}} \frac{\dot{V}_{kt}}{\eta_k} \right) \qquad \forall t \in \mathcal{T} \tag{B.11}$$

$$\dot{E}^h_{\mathcal{AC},t} \geq z_t \cdot \left(\frac{\dot{E}^c_t}{\eta^{min}_{\mathcal{AC}}} - \sum_{k \in \mathcal{AC}} \frac{\dot{V}_{kt}}{\eta_k} \right) \qquad \forall t \in \mathcal{T} \tag{B.12}$$

$$x_t + y_t + z_t = 1 \qquad \forall t \in \mathcal{T} \tag{B.13}$$

$$x, y, z \in \{0, 1\}^t . \tag{B.14}$$

Again, Golver's linearization (Glover, 1975) can be used to linearize the bilinear products with the binary variables x, y, and z.

B.3 Instances of demands

The thermal and electricity demands of instance 7 of the time series variation employing latin-hypercube sampling with $\pm 5\,\%$ are listed in Tab. B.1. The corresponding reliable solutions are discussed in detail in Section 5.2.2.

Table B.1: Thermal and electricity demands of instance 7 (for cooling on Site A and Site B); time step t_1 to t_5 are time-aggregated values; t_6 and t_7 represent minimal and maximal peak demands

demands	t_1	t_2	t_3	t_4	t_5	t_6	t_7
coolA in kW	1415	1747	2430	3129	4168	879	11797
coolB in kW	694	788	802	844	947	473	1705
heat in kW	4205	3293	2404	2312	2378	9708	1461
electricity in kW	5710	5181	5300	5444	5342	7663	7657

B.4 Costs for solutions based on varied demand data

Fig. B.2 shows the operational costs and investment costs for instance 6. Instance 6 represents the median of the nominal total annualized costs. Additionally, the ranges of all other instances are illustrated.

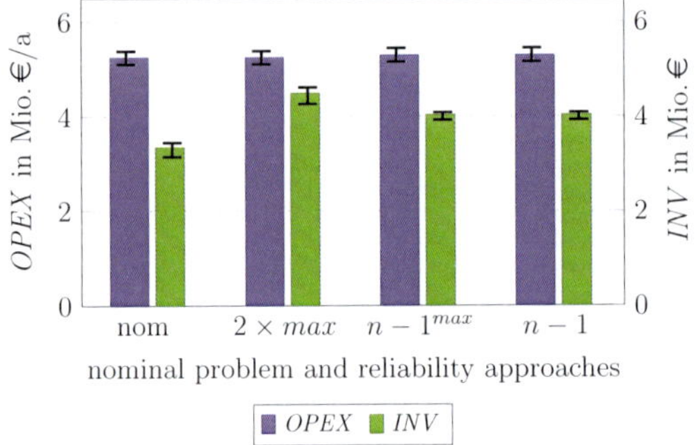

Figure B.2: Annualized operational costs *OPEX* and investment costs *INV* for the instance which represents the median of the nominal total annualized costs; error bars show the ranges in which all solutions for the other instances lie inside

Appendix C

Robustness

This appendix provides supplemental material for Chapter 6.

C.1 Problem formulation of the TRusT approach and of RoMO for robust and sustainable DESS

The RoMO problem formulation (Section 6.2.2) is based on a similar problem formulation as the TRusT problem formulation presented in Section 6.1.4. In the following, one problem formulation is presented for both approaches:

The design variables x^f include variables for the existence and for the installed thermal power $\dot{V}_{\mathrm{k}}^{N}$ of all components $k \in \mathcal{K}$. The design variables x^f do inherently not depend on time step $t \in \mathcal{T}$, in contrast to the operation variables x^s. The operation variables x^s include input and output energy flows for each time step t: Thermal and electrical input energy flows are denoted by $\dot{U}_{kt}$ for a component k, respectively. The variables $\dot{U}_t^{el,buy}$ describe the purchased electricity flow from the grid. Output energy flows are defined by $\dot{V}_{kt}$ and $\dot{V}_{kt}^{el}$ for a component k and by $\dot{V}_t^{el,sell}$ for sold electricity flow fed into the grid.

C.1.1 Objective functions

Only the set of considered objective functions differ for the TRusT approach and RoMO. The TRusT approach evaluates the trade-off between expected and worst-case costs by considering the nominal total annualized costs $TAC^{rob/nom}$ and the robust total annualized costs $TAC^{rob/rob}$ limited by α^{TAC}, both based on a robust efficient design:

$$\min \begin{pmatrix} TAC^{rob/nom} \\ \alpha^{TAC} \end{pmatrix}.$$

The nominal total annualized costs $TAC^{rob/nom}$ are defined by:

$$TAC^{rob/nom} := \sum_{t\in\mathcal{T}} \Big[\Delta\tau_t \Big(\hat{p}^{gas} \cdot \sum_{k\in\mathcal{B}\cup\mathcal{CHP}} \hat{\dot{U}}_{kt} + \hat{p}^{el,buy} \cdot \hat{\dot{U}}_t^{el,buy} - \hat{p}^{el,sell} \cdot \hat{\dot{V}}_t^{el,sell} \Big) \Big] + \sum_{k\in\mathcal{K}} \left(\frac{1}{PVF} + p_k^m \right) \cdot I_k$$

The auxiliary variable $\alpha^{TAC} \in \mathbb{R}$ restricts the robust objective function:

$$\begin{aligned} &\sum_{t\in\mathcal{T}} \Big[\Delta\tau_t \Big(\hat{p}^{gas}(1+\delta^{pg}) \cdot \sum_{k\in\mathcal{B}\cup\mathcal{CHP}} \dot{U}_{kt} \\ &\quad + \hat{p}^{el,buy}(1-\delta^{pe}) \cdot \dot{U}_t^{el,buy} - \hat{p}^{el,sell}(1-\delta^{pe}) \cdot \dot{V}_t^{el,sell} \Big) \Big] \\ &\quad + \sum_{k\in\mathcal{K}} \left(\frac{1}{PVF} + p_k^m \right) \cdot I_k \leq \alpha^{TAC} \\ &\sum_{t\in\mathcal{T}} \Big[\Delta\tau_t \Big(\hat{p}^{gas}(1+\delta^{pg}) \cdot \sum_{k\in\mathcal{B}\cup\mathcal{CHP}} \dot{U}_{kt} \\ &\quad + \hat{p}^{el,buy}(1+\delta^{pe}) \cdot \dot{U}_t^{el,buy} - \hat{p}^{el,sell}(1+\delta^{pe}) \cdot \dot{V}_t^{el,sell} \Big) \Big] \\ &\quad + \sum_{k\in\mathcal{K}} \left(\frac{1}{PVF} + p_k^m \right) \cdot I_k \leq \alpha^{TAC}. \end{aligned} \tag{C.1}$$

In RoMO, to design a sustainable energy supply system, we minimize the robust total annualized costs TAC^{rob} and robust global warming impact GWI^{rob}, i. e., the supremum of the uncertain total annualized costs $\sup_{\xi\in\mathcal{U}} TAC(x^f, x^s, \xi)$ and the supremum of the uncertain global warming impact $\sup_{\xi\in\mathcal{U}} GWI(x^s, \xi)$, respectively. For this reason, the uncertain total annualized costs $TAC(x^f, x^s, \xi)$ and the uncertain global warming impact $GWI(x^s, \xi)$ are limited by auxiliary variables $\alpha^{TAC} \in \mathbb{R}$ and $\alpha^{GWI} \in \mathbb{R}$, respectively, which are minimized:

$$\min \begin{pmatrix} \alpha^{TAC} \\ \alpha^{GWI} \end{pmatrix}.$$

Since the robust total annualized costs TAC^{rob} of RoMO corresponds to the robust total annualized costs $TAC^{rob/rob}$ of the TRusT approach, the same inequalities as introduced in Eq. (C.1) hold. The auxiliary variable α_{GWI} is the upper bound of the uncertain global warming impact $GWI(x^s, \xi)$ and also induces two additional constraints, as presented in Section 6.2.2:

$$\begin{aligned}
&\sum_{t\in\mathcal{T}} \Delta\tau_t \Bigg[\sum_{k\in\mathcal{B}\cup\mathcal{CHP}} \dot{U}_{kt} \cdot GWI^{gas} \\
&\qquad + \left(\dot{U}_t^{el,buy} - \dot{V}_t^{el,sell}\right) \cdot \left(\widehat{GWI}^{el} - \delta^{ge}\right)\Bigg] \leq \alpha^{GWI} \\
&\sum_{t\in\mathcal{T}} \Delta\tau_t \Bigg[\sum_{k\in\mathcal{B}\cup\mathcal{CHP}} \dot{U}_{kt} \cdot GWI^{gas} \\
&\qquad + \left(\dot{U}_t^{el,buy} - \dot{V}_t^{el,sell}\right) \cdot \left(\widehat{GWI}^{el} + \overline{\delta}^{ge}\right)\Bigg] \leq \alpha^{GWI}.
\end{aligned} \tag{C.2}$$

Instead of the total annualized costs other economic criteria could be chosen, e. g., the investment costs $INV = \sum_{k\in\mathcal{K}} I_k$ as in Appendix C.2.

For the combination of the TRusT approach and RoMO, as shown in Section 6.3, all three objective functions need to be minimized:

$$\min \begin{pmatrix} TAC^{rob/nom} \\ \alpha^{TAC} \\ \alpha^{GWI} \end{pmatrix}$$

employing the corresponding constraints (C.1) and (C.2).

C.1.2 Energy balances

The shared energy balances of the TRusT problem and the RoMO problem are based on the reformulations given in Section 6.1.4. For all $t \in \mathcal{T}$, maximal energy demands have to be covered at least:

$$\begin{aligned}
\sum_{k\in\mathcal{B}\cup\mathcal{CHP}} \dot{V}_{kt} - \sum_{k\in\mathcal{AC}} \dot{U}_{kt} &\geq \hat{\dot{E}}_t^{h} + \delta_t^{\dot{E}h} \\
\sum_{k\in\mathcal{AC}\cup\mathcal{CC}} \dot{V}_{kt} &\geq \hat{\dot{E}}_t^{c} + \delta_t^{\dot{E}c} \\
\sum_{k\in\mathcal{CHP}} \dot{V}_{kt}^{el} - \sum_{k\in\mathcal{CC}} \dot{U}_{kt}^{el} + \dot{U}_t^{el,buy} - \dot{V}_t^{el,sell} &\geq \hat{\dot{E}}_t^{el} + \delta_t^{\dot{E}e}.
\end{aligned}$$

In order to reduce overproduction, nominal and minimal demands have to be covered exactly for all time steps $t \in \mathcal{T}$ (Section 6.1.4). The additional operation variables $\widehat{x}^s$ and $\underline{x}^s$ allow to adapt the operation to the nominal and minimal demands. The lower bound variables $\underline{x}^s$ do not influence on the objective functions, whereas the nominal variables $\widehat{x}^s$ affect the nominal total annualized costs $TAC^{rob/nom}$ in the TRusT approach:

$$\sum_{k \in \mathcal{B} \cup \mathcal{CHP}} \widehat{\dot{V}}_{kt} - \sum_{k \in \mathcal{AC}} \widehat{\dot{U}}_{kt} = \widehat{\dot{E}}_t^h$$

$$\sum_{k \in \mathcal{AC} \cup \mathcal{CC}} \widehat{\dot{V}}_{kt} = \widehat{\dot{E}}_t^c$$

$$\sum_{k \in \mathcal{CHP}} \widehat{\dot{V}}_{kt}^{el} - \sum_{k \in \mathcal{CC}} \widehat{\dot{U}}_{kt}^{el} + \widehat{\dot{U}}_t^{el,buy} - \widehat{\dot{V}}_t^{el,sell} = \widehat{\dot{E}}_t^{el}$$

$$\sum_{k \in \mathcal{B} \cup \mathcal{CHP}} \underline{\dot{V}}_{kt} - \sum_{k \in \mathcal{AC}} \underline{\dot{U}}_{kt} = \widehat{\dot{E}}_t^h - \delta_t^{\dot{E}h}$$

$$\sum_{k \in \mathcal{AC} \cup \mathcal{CC}} \underline{\dot{V}}_{kt} = \widehat{\dot{E}}_t^c - \delta_t^{\dot{E}c}$$

$$\sum_{k \in \mathcal{CHP}} \underline{\dot{V}}_{kt}^{el} - \sum_{k \in \mathcal{CC}} \underline{\dot{U}}_{kt}^{el} + \underline{\dot{U}}_t^{el,buy} - \underline{\dot{V}}_t^{el,sell} = \widehat{\dot{E}}_t^{el} - \delta_t^{\dot{E}e}.$$

The output energy flows $\dot{V}_{kt}$, $\widehat{\dot{V}}_{kt}$, and $\underline{\dot{V}}_{kt}$ of a component are limited by the installed thermal power $\dot{V}_k^N$ and by the minimal part-load given as fraction of the installed thermal power using the factor μ^{min}:

$$\begin{aligned}
\mu^{min} \cdot \dot{V}_k^N &\leq \dot{V}_{kt} \leq \dot{V}_k^N && \forall k \in \mathcal{K}, t \in \mathcal{T} \\
\mu^{min} \cdot \dot{V}_k^N &\leq \widehat{\dot{V}}_{kt} \leq \dot{V}_k^N && \forall k \in \mathcal{K}, t \in \mathcal{T} \\
\mu^{min} \cdot \dot{V}_k^N &\leq \underline{\dot{V}}_{kt} \leq \dot{V}_k^N && \forall k \in \mathcal{K}, t \in \mathcal{T}.
\end{aligned}$$

Besides the constraints given in this section, constraints regarding the efficiencies of components as well as constraints for linearization of the investment costs I_k of the components need to be employed (see Appendix A.2).

C.2 Global warming impact versus investment costs

In this section, we investigate the bi-objective problem for the investment costs INV and the global warming impact $GWI(\xi)$ to determine a sustainable DESS.

The investment costs INV are defined by:

$$INV = \sum_{k \in \mathcal{K}} I_k \,. \tag{C.3}$$

Since the investment costs are not effected by the uncertainties, the objective-wise RoMO problem further simplifies as the following example shows.

Example C.1 *The problem further simplifies if only one objective function contains uncertain parameters. For example, when we consider the investment costs INV as economic criterion, we assume all parameters influencing the investment costs INV to be known with perfect foresight. Thus, there is no dependence on the scenario ξ. As a result, the range of possible objective function values forms only a vertical line (Fig. C.1).*

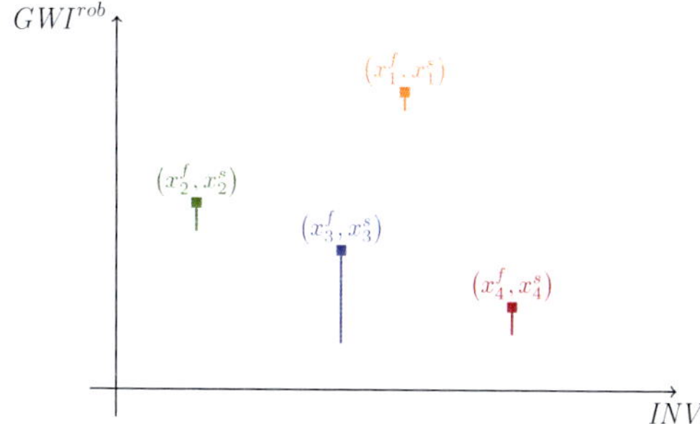

Figure C.1: The range of objective function values of the robust global warming impact GWI^{rob} and the investment costs INV are presented for the 4 robust feasible solutions $(x^f, x^s) \in \mathcal{X}$. Possible objective function values form vertical lines. The worst-case values $\left(INV(x^f),\ GWI\,(x^s, \xi^{wc})\right)^T$ for $(x^f, x^s) \in \mathcal{X}$ are highlighted with small squares. The solutions $(x_2^f, x_2^s), (x_3^f, x_3^s)$, and (x_4^f, x_4^s) are robust efficient.

The inner subproblem $\left(INV(x^f),\ \sup_{\xi \in \mathcal{U}} GWI\,(x^s, \xi)\right)^T$ results in one pair of values $\left(INV(x^f),\ GWI\,(x^s, \xi^{wc})\right)^T$ for each robust feasible solution $(x^f, x^s) \in \mathcal{X}$, since the occurrence of uncertainties in only one objective function is a special case of the presented objective-wise uncertainty. Bi-objective problems with only one uncertain objective function have been analyzed in detail by Kuhn et al. (2016).

C.2.1 Results of the case study

Results of the case study are presented regarding the global warming impact versus the investment costs.

Efficient design options with perfect foresight

In this section, we analyze the efficient solutions of the nominal bi-objective optimization problem of designing a sustainable energy supply system of an industrial park. For this purpose, perfect foresight is assumed employing only the nominal scenario. The nominal efficient solutions help to evaluate the robust efficient solutions in the following section. The nominal optimization problem comprises up to 41 609 continuous variables and up to 2 596 binary variables. The number of constraints is about 45 700. Generating the whole Pareto front with 100 solutions needs about 14.4 minutes.

Fig. C.2 shows the Pareto front of the nominal problem with investment costs INV (Eq. C.3) and the nominal global warming impact GWI^{nom} (Eq. 4.2) as objective functions.

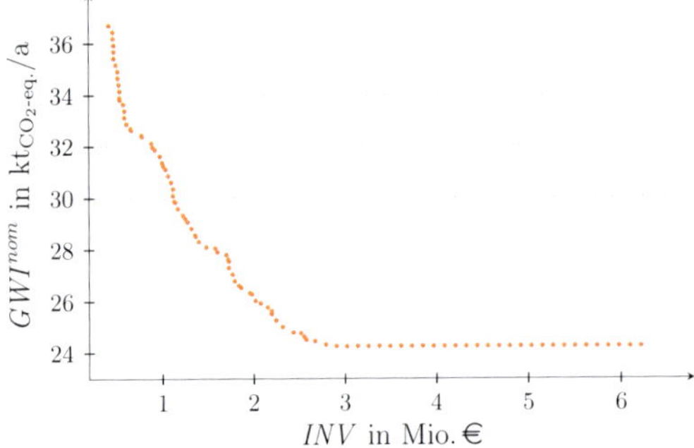

Figure C.2: Pareto front of the nominal bi-objective optimization problem regarding investment costs INV and global warming impact GWI^{nom}

The maximal global warming impact on the Pareto front is $GWI^{max} = 36.7\, kt_{CO_2\text{-eq.}}/a$ resulting from the design with minimal investment costs of $INV^{min} = 0.4$ Mio. €. Minimizing the global warming impact yields

$GWI^{min} = 24.3\,\mathrm{kt_{CO_2\text{-eq.}}}/\mathrm{a}$ with investment costs of $INV^{max} = 6.2\,\mathrm{Mio.}\,€$. The Pareto front in Fig. C.2 shows that 70 % of the maximal global warming impact reduction is already achieved at investment costs of about 1.5 Mio. €.

In the Pareto front, small kinks can be observed where the slope changes. The *steep parts* mark parts of the Pareto front where one objective function improves significantly while the second objective function only worsen marginally. In the *flat parts*, small improvement of the first objective function would lead to a significant worsening of the second objective function. E. g., at 0.9 Mio. € and 1.7 Mio. €, the investments increase without a notable benefit in the global warming impact. These flat parts are caused by an additionally installed combined heat and power engine (Fig. C.3). Combined heat and power engines are more expensive than other equipment, i. e., boilers, absorption chillers, and compression chillers. Thus, at low investments, few combined heat and power engines are installed. All newly installed components are shown in Fig. C.3.

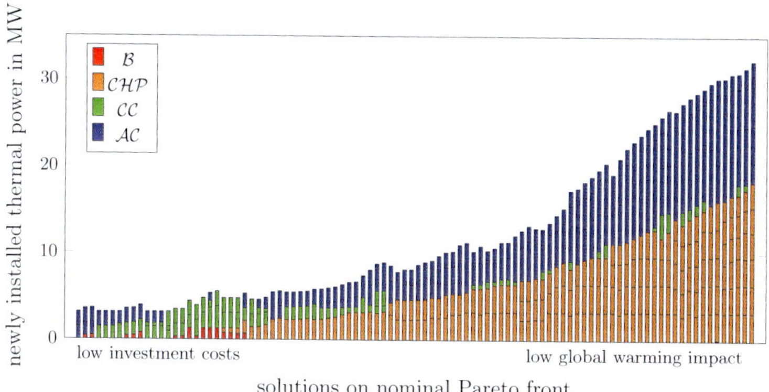

Figure C.3: Nominal efficient design options for newly installed components: $\mathcal{B}$ boilers, $\mathcal{CHP}$ combined heat and power engines, $\mathcal{CC}$ compression chillers, and $\mathcal{AC}$ absorption chillers.

At low investment costs, absorption chillers are the first choice to cover the new cooling demand, since their investment costs are lower than the investments for compression chillers. For slightly increasing investments, a boiler supplies heating until the first combined heat and power engine is large enough to supply sufficient heating. Compression chillers support the absorption chillers until the second combined heat and power engine is implemented. For further increasing investment costs, the cooling demands are mostly covered by absorption chillers using heat from the combined heat and power engines. The same effect of switching from compression chillers to absorption chillers can be observed when installing the third and fourth combined heat and power engine.

Robust efficient design options

RoMO is employed to determine robust sustainable designs regarding investment costs INV and the robust global warming impact GWI^{rob}. We consider uncertainties in the energy demands $\dot{E}_t$ and in the specific global warming impact of the electricity mix GWI^{el} of the grid (Section 4.2). Thereby, we enable designing cost-reasonable DESS with a low robust global warming impact GWI^{rob}.

The RoMO problem formulation contains around 32 300 continuous variables and about 2 100 binary variables for each optimization problem. The number of constraints varies around 36 000. The computational time for all 100 solutions lies with about 14.2 minutes in a similar range as the computational time for the nominal optimization problem.

Fig. C.4 illustrates the effect of the uncertainties on the Pareto front. The nominal Pareto front has been analyzed in the previous section and is presented in Fig. C.4 only for comparison.

Considering only uncertainty of the specific global warming impact of the electricity mix GWI^{el} deforms the Pareto front (Fig. C.4): The minimal global warming impact is about 26.2 $\mathrm{kt_{CO_2\text{-eq.}}/a}$. For investments above 2.1 Mio. €, the robust Pareto front, based on an uncertain specific global warming impact $\widetilde{GWI}^{el}$, shows only a marginal decrease of the robust global warming impact $GWI^{rob}_{GWI^{el}}$.

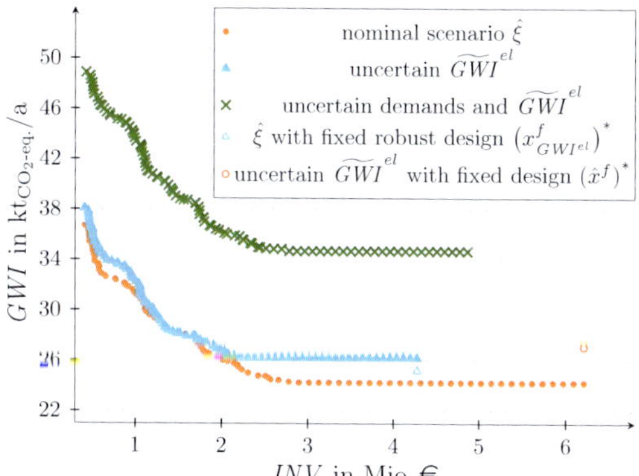

Figure C.4: Pareto fronts regarding investment costs INV and global warming impact GWI with certain and uncertain parameters: The nominal Pareto front is presented only for comparison (orange dots; see also Fig. C.2); the Pareto front considering an uncertain specific global warming impact of the electricity mix $\widetilde{GWI}^{el}$ is shown by filled triangles; if demands are also uncertain, the robust Pareto front is marked by dark green crosses; for open symbols, uncertainties in the demands are not taken into account: the unfilled triangle marks the optimal solution for the nominal scenario employing the robust optimal design $\left(x^f_{GWI^{el}}\right)^*$ from minimizing the robust global warming impact $GWI^{rob}_{GWI^{el}}$; vice versa, the unfilled circle marks the nominal optimal design $(\hat{x}^f)^*$ employed for minimizing the robust global warming impact GWI^{rob}.

As discussed in the previous section, the minimal global warming impact in the nominal scenario is $GWI^{min} = 24.3\,kt_{CO_2\text{-eq.}}/a$ (Fig. C.4, filled dots, right anchor point). Employing the corresponding nominal optimal design $(\hat{x}^f)^*$ and allowing for uncertainty of the specific global warming impact GWI^{el} leads to global warming impact of $27.1\,kt_{CO_2\text{-eq.}}/a$ in the worst-case scenario $\xi^{wc}_{GWI^{el}}$ (Fig. C.4, unfilled dot). In contrast, the minimal robust global warming impact corresponds only to $26.2\,kt_{CO_2\text{-eq.}}/a$ (Fig. C.4, right anchor point of filled triangles). Thus, the nominal optimal solution for the minimal global warming impact becomes suboptimal, if uncertainty of the specific global warming impact $\widetilde{GWI}^{el}$ is regarded. This occurs even though the investment costs are 31 % lower for the robust optimal solution. In contrast, employing the robust optimal design $\left(x^f_{GWI^{el}}\right)^*$ for nominal operation leads to a global warming impact of 25.2 $kt_{CO_2\text{-eq.}}/a$. Compared to the minimal global warming impact GWI^{min} of the best nominal design $(\hat{x}^f)^*$, the increase corresponds to only 3.9 %.

Taking also uncertain demands into account shifts the Pareto front to a higher level of global warming impact. The global warming impact increases because the energy supply system must cover higher demands in the worst case for which the objective function is evaluated. The shift of the Pareto front corresponds to about 9 $kt_{CO_2\text{-eq.}}/a$. The nominal design options would become infeasible if allowing for uncertain demands.

The robust Pareto front (Fig. C.4, dark green crosses) shows that investments below 2.4 Mio. € improve the global warming impact significantly: At low investments, a small increase in the investment costs INV strongly reduces the robust global warming impact GWI^{rob}, e. g., investing 1.1 Mio. € instead of 0.9 Mio. € leads to savings of 3.8 $kt_{CO_2\text{-eq.}}/a$. However, investments above 2.4 Mio. € again provide very little additional improvement of the robust global warming impact GWI^{rob}. The small improvement of the robust global warming impact GWI^{rob} is related to the employed robustness method: Strictly robust optimization inherently considers the worst case. In the considered problem, the worst case depends on whether electricity is purchased from the grid or fed in. When electricity is purchased, the worst case is the upper bound of the specific global warming impact $\widetilde{GWI}^{el}$. As a result, the system tends to purchase less electricity, since this would increase the global warming impact significantly. In contrast, when feeding in electricity, the lower bound of the specific global warming impact $\widetilde{GWI}^{el}$ is decisive, since it leads to a smaller avoided burden. Thus, the system favors on-site generated electricity by combined heat and power engines without feeding in electricity. As a result, the system does not use the connection to the electricity grid to purchase or to feed in electricity. Hence, the energy system tends to be electricity-autonomous for high investments. However, for small investment costs, the system purchases electricity from the grid, since the already existing combined heat and power engine has a poor efficiency. Thus, electricity from the grid is favored even though the upper bound of the specific global warming impact $\widetilde{GWI}^{el}$ needs to be taken into account.

Similar to the nominal Pareto front, the robust Pareto front also contains kinks. The kinks in the robust Pareto front occur due to the same reason: While reducing the robust global warming impact GWI^{rob} slightly, the investment costs INV increase notably along the robust Pareto front when an additional combined heat and power engine is installed (Fig. C.5).

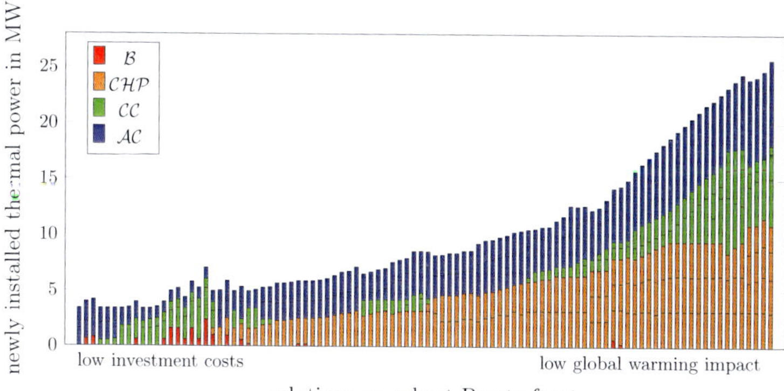

Figure C.5: Robust efficient design options for newly installed components: $\mathcal{B}$ boilers, $\mathcal{CHP}$ combined heat and power engines, $\mathcal{CC}$ compression chiller, and $\mathcal{AC}$ absorption chillers.

As also observed in Section 6.2.3, the robust efficient solutions show similar characteristics as the nominal efficient solutions (Fig. C.5): Combined heat and power engines and absorption chillers replace boilers and compression chillers when the investment costs increase. However, the uncertain specific global warming impact of the electricity mix $\widetilde{GWI}^{el}$ leads to significant differences for the robust efficient solutions: In total, the robust efficient solutions contain less thermal power of combined heat and power engines than the nominal efficient solutions (compare Fig. C.5 with Fig. C.3). The reason is that for investment costs above 2.4 Mio. € the connection to the electricity grid is not advantageous when regarding an uncertain specific global warming impact $\widetilde{GWI}^{el}$; and thus, a positive effect on the global warming impact by selling electricity cannot be guaranteed. As a result, the electricity provided by the combined heat and power engines is only used on site to supply compression chillers and electricity demands, and is not fed into the electricity grid.

C.3 Further results of the case study combining the TRusT approach and RoMO

Further results of the combination of the TRusT approach and RoMO (Section 6.3) are presented.

Fig. C.6 shows 3 perspectives of the Pareto front of the combination of the TRusT approach and RoMO regarding the nominal total annualized costs $TAC^{rob/nom}$, the robust total annualized costs TAC^{rob}, and the robust global warming impact GWI^{rob}.

(a)

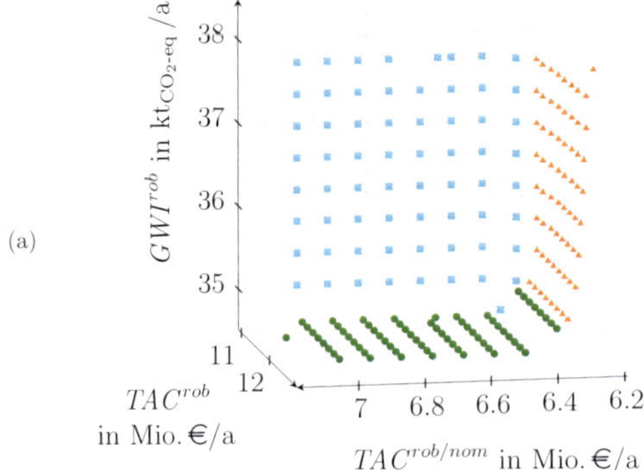

Figure C.6: 3-dimensional Pareto front regarding the nominal total annualized costs $TAC^{rob/nom}$, the robust total annualized costs TAC^{rob}, and the robust global warming impact GWI^{rob}; the Pareto front is shown from 3 sides for improved visualization ((a), (b), and (c)); orange triangles are computed minimizing the nominal total annualized costs $TAC^{rob/nom}$, blue squares minimizing the robust total annualized costs TAC^{rob}, and green dots minimizing the robust global warming impact GWI^{rob}

(b) 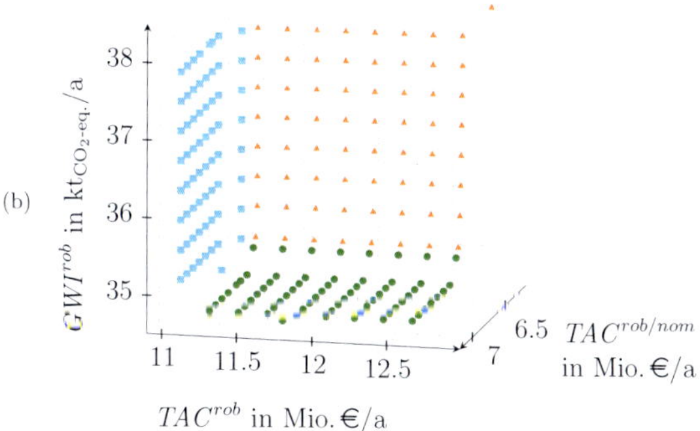

(c) 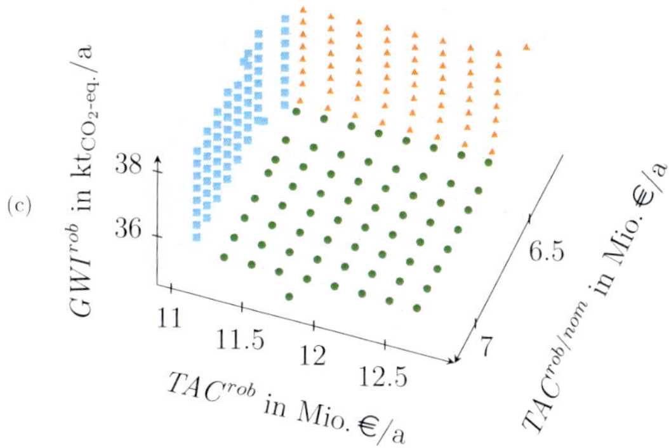

The designs on the Pareto front are shown (Fig. C.7) corresponding to the solutions obtained by minimizing nominal total annualized costs $TAC^{rob/nom}$ and by minimizing the robust total annualized costs TAC^{rob} (marked by orange triangles and blue squared in Fig. C.6, respectively). In Fig. C.7, no trends in the system design along the Pareto front can be observed.

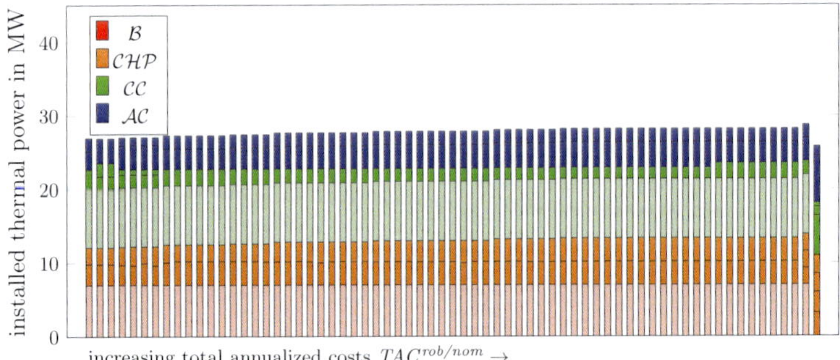

(a) Robust efficient design (structure and sizing) of installed components minimizing the nominal total annualized costs $TAC^{rob/nom}$ (orange triangles in Fig. 6.13)

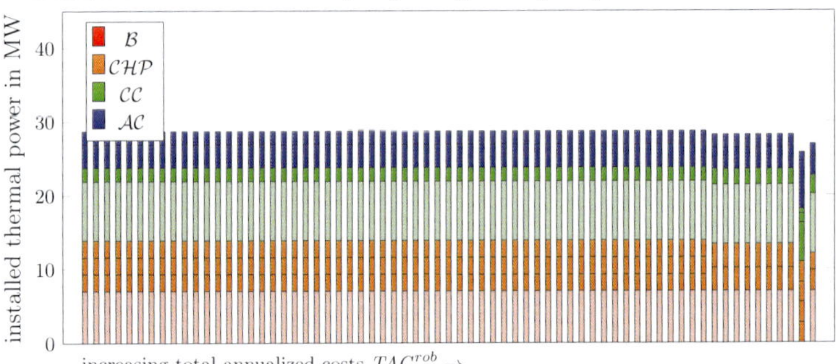

(b) Robust efficient design (structure and sizing) of installed components minimizing the robust total annualized costs TAC^{rob} (blue squares in Fig. 6.13)

Figure C.7: Design of efficient solutions computed minimizing the nominal total annualized costs $TAC^{rob/nom}$ and the robust total annualized costs TAC^{rob}; $\mathcal{B}$ boilers, $\mathcal{CHP}$ combined heat and power engines, $\mathcal{CC}$ compression chiller, and $\mathcal{AC}$ absorption chillers; components shown in light colors remain from the already existing components.

Appendix D

Design selection

This appendix provides supplemental material for Chapter 7.

D.1 Problem formulation of the flex-hand approach

In the following, we provide the problem formulation of the flex-hand optimization for DESS. We consider the total annualized costs TAC and the global warming impact GWI as criteria (Section 4.1.1).

$$
\begin{aligned}
\min\ & \overline{\varepsilon} \\
\text{s.t.}\ & \text{objective functions} \\
& \overline{TAC}\left(\dot{U}, \dot{U}^{el,buy}, \dot{V}^{el,sell}, \gamma, \dot{V}^{N}, j\right) - \overline{TAC}^{*}(j) \leq \overline{\varepsilon} && \forall j \in [N] \\
& \overline{GWI}\left(\dot{U}, \dot{U}^{el,buy}, \dot{V}^{el,sell}, j\right) - \overline{GWI}^{*}(j) \leq \overline{\varepsilon} && \forall j \in [N] \\
& \text{energy balances} \\
& \sum_{k \in \mathcal{B} \cup \mathcal{CHP}} \dot{V}_{kt}(j) - \sum_{k \in \mathcal{AC}} \dot{U}_{kt}(j) = \dot{E}_{t}^{heat} && \forall t \in \mathcal{T}, j \in [N] \\
& \sum_{k \in \mathcal{AC} \cup \mathcal{CC}} \dot{V}_{kt}(j) = \dot{E}_{t}^{cool} && \forall t \in \mathcal{T}, j \in [N] \\
& \sum_{k \in \mathcal{CHP}} \dot{V}_{kt}^{el}(j) - \sum_{k \in \mathcal{CC}} \dot{U}_{kt}(j) \\
& \quad + \dot{U}_{t}^{el,buy}(j) - \dot{V}_{t}^{el,sell}(j) = \dot{E}_{t}^{el} && \forall t \in \mathcal{T}, j \in [N]
\end{aligned}
$$

linearization of investment costs

$$\sum_{h\in[H]} \gamma_{kh} \leq 1 \qquad \forall k \in \mathcal{K}$$

$$\gamma_{kh} \cdot \dot{V}_{kh}^{N,lb} \leq \dot{V}_{kh}^{N} \leq \gamma_{kh} \cdot \dot{V}_{kh+1}^{N,lb} \qquad \forall k \in \mathcal{K}, h \in [H]$$

further constraints and variables

$$\mu^{min} \cdot \sum_{h\in[H]} \dot{V}_{kh}^{N} \leq \dot{V}_{kt}(j) \leq \sum_{h\in[H]} \dot{V}_{kh}^{N} \qquad \forall k \in \mathcal{K}, t \in \mathcal{T}, j \in [N]$$

$$\dot{V}_{kt}(j) = \eta_k \cdot \dot{U}_{kt}(j) \qquad \forall k \in \mathcal{K}, t \in \mathcal{T}, j \in [N]$$

$$\dot{V}_{kt}^{el}(j) = \eta_k^{tot} \cdot \dot{U}_{kt}(j) - \dot{V}_{kt}(j) \qquad \forall k \in \mathcal{CHP}, t \in \mathcal{T}, j \in [N]$$

$$\dot{U}^{el,buy}(j), \dot{V}^{el,sell}(j), \dot{V}^{el}(j), \dot{U}(j), \dot{V}(j) \in \mathbb{R}_+^{|\mathcal{K}|\times T} \qquad \forall j \in [N]$$

$$\gamma \in \{0,1\}^{|\mathcal{K}|\times H}, \dot{V}^{N} \subset \mathbb{R}_+^{|\mathcal{K}|\times H}$$

$$\bar{\varepsilon} \subset \mathbb{R}_+$$

In the problem formulation, all second-stage variables depend on the index j, to be able to adapt operation for each point on the flex-hand Pareto front. The total annualized costs TAC and the global warming impact GWI are defined by

$$\begin{aligned} TAC&\left(\dot{U}, \dot{U}^{el,buy}, \dot{V}^{el,sell}, \gamma, \dot{V}^{N}, j\right) \\ &= \sum_{t\in\mathcal{T}} \left[\Delta\tau_t \left(p^{gas} \cdot \sum_{k\in\mathcal{B}\cup\mathcal{CHP}} \dot{U}_{kt}(j) + p^{el,buy} \cdot \dot{U}_t^{el,buy}(j) - p^{el,sell} \cdot \dot{V}_t^{el,sell}(j)\right)\right] \\ &\qquad + \sum_{k\in\mathcal{K}} \left(\frac{1}{PVF} + p_k^m\right) \cdot \underbrace{\sum_{h\in[H]} \left[\gamma_{kh} \cdot \kappa_{kh} + m_{kh} \cdot \left(\dot{V}_{kh}^{N} - \gamma_{kh}\dot{V}_{kh}^{N,lb}\right)\right]}_{=:I_k} \end{aligned}$$

$$\begin{aligned} GWI&\left(\dot{U}, \dot{U}^{el,buy}, \dot{V}^{el,sell}, j\right) \\ &= \sum_{t\in\mathcal{T}} \Delta\tau_t \left[\sum_{k\in\mathcal{B}\cup\mathcal{CHP}} \dot{U}_{kt}(j) \cdot GWI^{gas} + \left(\dot{U}_t^{el,buy}(j) - \dot{V}_t^{el,sell}(j)\right) \cdot GWI^{el}\right]. \end{aligned}$$

Bars above the total annualized costs TAC and the global warming impact GWI in the optimization problem denote the normalization of the objective values. Objective values on the normalized ideal Pareto fronts are denoted by $\left(\overline{TAC}^*(j), \overline{GWI}^*(j)\right)$ for each point $j \in [N]$.

D.2 Application of the flex-hand approach to single-stage optimization problems

Applying the flex-hand approach to a single-stage optimization problem, the approach minimizes the objective-wise distance to the ideal point (Fig. D.1).

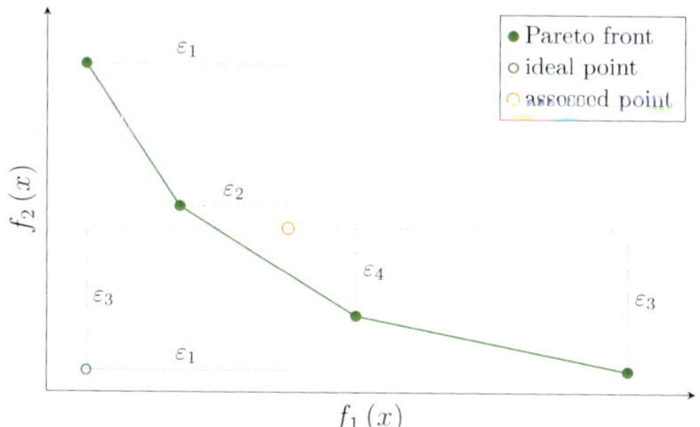

Figure D.1: Application of the flex-hand approach to a single-stage optimization problem; as there is no second stage, only one point is assessed; ε_1, ε_2, ε_3, and ε_4 represent the differences of objective values with $\varepsilon_1 > \varepsilon_2$ and $\varepsilon_3 > \varepsilon_4$; lines are included to guide reader's eyes

As there is no second stage, the fixed first-stage Pareto front degenerates to only one point. The inner points on the Pareto front always lead to smaller differences of the objective values than the anchor points (in Fig. D.1: $\varepsilon_1 > \varepsilon_2$ and $\varepsilon_3 > \varepsilon_4$). Thus, only the anchor points are decisive. Since the differences of the anchor points to the assessed point are identical to the differences regarding the ideal point, the flex-hand approach reduces to an approach minimizing the objective-wise distance to the ideal point.

D.3 Scenario-based design selection

Instead of employing RoMO, uncertainties can directly be included in the flex-hand approach. For including uncertainties, the flex-hand approach is extended by considering uncertain input parameters based on scenarios.

D.3.1 The scenario-based flex-hand approach

In optimization of DESS, decisions are based on input parameters which are inherently uncertain. Thus, we extend the proposed flex-hand approach for problems comprising uncertain parameters in the objective functions and constraints, and introduce the *scenario-based flex-hand approach*. As the flex-hand approach, the scenario-based flex-hand approach can also be applied to any other two-stage multi-objective problem where the first-stage needs to be determined in advance. The scenario-based flex-hand approach provides an alternative way to enable robustness. While RoMO (Section 6.2) considers all scenarios simultaneously, the scenario-based approach adapts operation to each scenario separately; thus, a discrete uncertainty set $\mathcal{U}$ is necessary. For each scenario, we compare the ideal Pareto front for the current scenario to a fixed first-stage Pareto front which is based on one fixed set of first-stage variables for *all* scenarios simultaneously. We then minimize the distance between the Pareto fronts in the worst case and thereby find the robust optimal first-stage solution.

For this purpose, we calculate the ideal Pareto front $\mathcal{P}^*(\xi)$ in each scenario $\xi \in \mathcal{U}$ separately. The number of elements in $\mathcal{P}^*(\xi)$ is denoted by $N(\xi)$. The objectives are parametrized also through scenarios $\xi \in \mathcal{U}$, i. e., we use $f_i\big(x^f, x^s(\xi), \xi\big)$. Again, we normalize objectives which we denote by $\overline{f}_i\big(x^f, x^s(\xi), \xi\big)$, for each scenario ξ. To minimize the worst-case distance between the ideal Pareto front and the fixed first-stage Pareto front, we minimize the maximum value of the ε-indicator (Eq. (7.1)) over all $\xi \in \mathcal{U}$:

$$
\begin{aligned}
\min\ & \overline{\varepsilon} \\
\text{s.t. } & \overline{f}_i\big(x^f, x^s(\xi, j), \xi\big) - \overline{p}_i^*(\xi, j) \leq \overline{\varepsilon} && \forall \xi \in \mathcal{U}, i \in [L], j \in [N(\xi)] \\
& x^f \in \mathcal{X}^f \\
& x^s(\xi, j) \in \mathcal{X}^s(x^f, \xi) && \forall \xi \in \mathcal{U}, j \in [N(\xi)],
\end{aligned}
$$

where $\overline{p}_i^*(\xi, j)$ is the normalized objective function value in scenario ξ of point j on the ideal Pareto front $\mathcal{P}^*(\xi)$. $\mathcal{X}^s(x^f, \xi)$ is the set of feasible second-stage variables for a given first-stage solution x^f in scenario ξ. The optimal first-stage solution $(x^f)^*$ is the selected *robust flex-hand solution*.

An example is given in Figure D.2. The ideal Pareto front is calculated for each of the three scenarios $\{\xi_1, \xi_2, \xi_3\}$ separately. The robust fixed first-stage Pareto fronts are based on one fixed set of first-stage variables for all scenarios; however, the corresponding robust fixed first-stage Pareto fronts are calculated for each scenario separately by adapting the second-stage variables x^s.

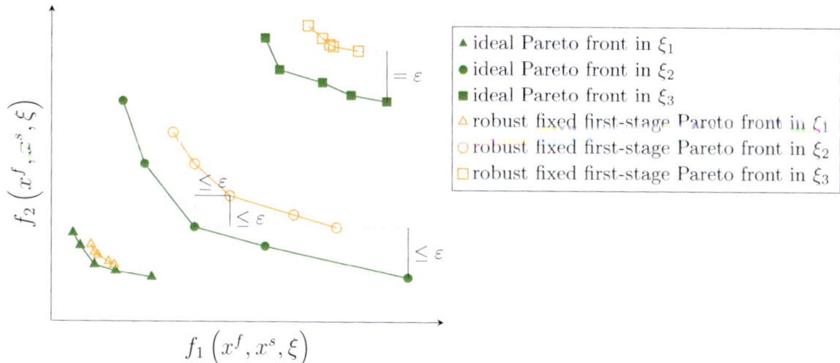

Figure D.2: Idea of the scenario-based flex-hand approach: For each scenario separately, the ideal Pareto front (dark green filled marks) and an arbitrary robust fixed first-stage Pareto front (orange unfilled marks) are compared; triangles, circles, and squares represent scenario ξ_1, ξ_2, and ξ_3, respectively; the distance ε is calculated regarding all scenarios; Pareto fronts are presented without normalization; lines are included to guide reader's eyes

D.3.2 Model formulation of the scenario-based flex-hand approach for DESS optimization

In the following, we provide the problem formulation of the scenario-based flex-hand optimization for the DESS. Uncertainties are regarded for prices for purchasing gas $\widetilde{p}^{gas}$ and electricity $\widetilde{p}^{el,buy}$ as well as for selling electricity $\widetilde{p}^{el,sell}$ (Section 4.2). Furthermore, the specific global warming impact of the electricity mix of the grid $\widetilde{GWI}^{el}$ is assumed to be uncertain as well. In the constraints, the energy balances are affected by uncertain energy demands $\widetilde{\dot{E}}^{heat}$, $\widetilde{\dot{E}}^{cool}$, and $\widetilde{\dot{E}}^{el}$.

$$
\begin{aligned}
\min\ & \bar{\varepsilon} \\
\text{s. t.}\ & \text{objective functions} \\
& \overline{TAC}\left(\dot{U}, \dot{U}^{el,buy}, \dot{V}^{el,sell}, \gamma, \dot{V}^{N}, \xi, j\right) \\
& \qquad\qquad - \overline{TAC}^{*}(\xi, j) \leq \bar{\varepsilon} && \forall \xi \in \mathcal{U}, j \in [N(\xi)] \\
& \overline{GWI}\left(\dot{U}, \dot{U}^{el,buy}, \dot{V}^{el,sell}, \xi, j\right) \\
& \qquad\qquad - \overline{GWI}^{*}(\xi, j) \leq \bar{\varepsilon} && \forall \xi \in \mathcal{U}, j \in [N(\xi)] \\
& \text{scenario-based energy balances} \\
& \sum_{k \in \mathcal{B} \cup \mathcal{CHP}} \dot{V}_{kt}(\xi, j) - \sum_{k \in \mathcal{AC}} \dot{U}_{kt}(\xi, j) = \tilde{\dot{E}}_t^{heat} && \forall t \in \mathcal{T}, \xi \in \mathcal{U}, j \in [N(\xi)] \\
& \sum_{k \in \mathcal{AC} \cup \mathcal{CC}} \dot{V}_{kt}(\xi, j) = \tilde{\dot{E}}_t^{cool} && \forall t \in \mathcal{T}, \xi \in \mathcal{U}, j \in [N(\xi)] \\
& \sum_{k \in \mathcal{CHP}} \dot{V}_{kt}^{el}(\xi, j) - \sum_{k \in \mathcal{CC}} \dot{U}_{kt}(\xi, j) \\
& \qquad + \dot{U}_t^{el,buy}(\xi, j) - \dot{V}_t^{el,sell}(\xi, j) = \tilde{\dot{E}}_t^{el} && \forall t \in \mathcal{T}, \xi \in \mathcal{U}, j \in [N(\xi)] \\
& \text{linearization of investment costs} \\
& \sum_{h \in [H]} \gamma_{kh} \leq 1 && \forall k \in \mathcal{K} \\
& \gamma_{kh} \cdot \dot{V}_{kh}^{N,lb} \leq \dot{V}_{kh}^{N} \leq \gamma_{kh} \cdot \dot{V}_{kh+1}^{N,lb} && \forall k \in \mathcal{K}, h \in [H] \\
& \text{further constraints and variables} \\
& \mu^{min} \cdot \sum_{h \in [H]} \dot{V}_{kh}^{N} \leq \dot{V}_{kt}(\xi, j) \leq \sum_{h \in [H]} \dot{V}_{kh}^{N} && \forall k \in \mathcal{K}, t \in \mathcal{T}, \xi \in \mathcal{U}, j \in [N(\xi)] \\
& \dot{V}_{kt}(\xi, j) = \eta_k \cdot \dot{U}_{kt}(\xi, j) && \forall k \in \mathcal{K}, t \in \mathcal{T}, \xi \in \mathcal{U}, j \in [N(\xi)] \\
& \dot{V}_{kt}^{el}(\xi, j) = \eta_k^{tot} \cdot \dot{U}_{kt}(\xi, j) - \dot{V}_{kt}(\xi, j) && \forall k \in \mathcal{CHP}, t \in \mathcal{T}, \\
& && \forall \xi \in \mathcal{U}, j \in [N(\xi)] \\
& \dot{U}^{el,buy}(\xi, j), \dot{V}^{el,sell}(\xi, j), \dot{V}^{el}(\xi, j), \in \mathbb{R}_{+}^{|\mathcal{K}| \times T} && \forall \xi \in \mathcal{U}, j \in [N(\xi)] \\
& \dot{U}(\xi, j), \dot{V}(\xi, j) \in \mathbb{R}_{+}^{|\mathcal{K}| \times T} && \forall \xi \in \mathcal{U}, j \in [N(\xi)] \\
& \gamma \in \{0, 1\}^{|\mathcal{K}| \times H}, \dot{V}^{N} \in \mathbb{R}_{+}^{|\mathcal{K}| \times H} \\
& \bar{\varepsilon} \in \mathbb{R}_{+}
\end{aligned}
$$

In the problem formulation, all operational variables depend on the index j and the occurring scenario ξ, to be able to adapt the supplied energy to each point on the Pareto front for each scenario. The total annualized costs TAC and the global warming impact GWI are defined by

$$
\begin{aligned}
&TAC\left(\dot{U}, \dot{U}^{el,buy}, \dot{V}^{el,sell}, \gamma, \dot{V}^{N}, \xi, j\right) \\
&= \sum_{t\in\mathcal{T}} \Bigg[\Delta\tau_t\bigg(\widetilde{p}^{gas} \cdot \sum_{k\in\mathcal{B}\cup\mathcal{CHP}} \dot{U}_{kt}(\xi,j) + \widetilde{p}^{el,buy} \cdot \dot{U}_t^{el,buy}(\xi,j) \\
&\qquad\qquad\qquad\qquad - \widetilde{p}^{el,sell} \cdot \dot{V}_t^{el,sell}(\xi,j)\bigg)\Bigg] \\
&\qquad + \sum_{k\in\mathcal{K}} \left(\frac{1}{PVF} + p_k^{m}\right) \cdot \underbrace{\sum_{h\in[H]} \left[\gamma_{kh} \cdot \kappa_{kh} + m_{kh} \cdot \left(\dot{V}_{kh}^{N} - \gamma_{kh}\dot{V}_{kh}^{N,lb}\right)\right]}_{=:I_k} \\
&GWI\left(\dot{U}, \dot{U}^{el,buy}, \dot{V}^{el,sell}, \xi, j\right) \\
&= \sum_{t\in\mathcal{T}} \Delta\tau_t \left[\sum_{k\in\mathcal{B}\cup\mathcal{CHP}} \dot{U}_{kt}(\xi,j) \cdot GWI^{gas} + \left(\dot{U}_t^{el,buy}(\xi,j) - \dot{V}_t^{el,sell}(\xi,j)\right) \cdot \widetilde{GWI}^{el}\right].
\end{aligned}
$$

Bars above the total annualized costs TAC and the global warming impact GWI in the optimization problem denote the normalization of the objective values. Objective values on the normalized ideal Pareto fronts are denoted by $\left(\overline{TAC}^*(\xi,j), \overline{GWI}^*(\xi,j)\right)$ for each point $j \in [N(\xi)]$ and each scenario ξ of the uncertainty set $\mathcal{U}$.

D.3.3 Results of the case study

We now apply the scenario-based flex-hand approach to the proposed case study taking uncertainties into account. The uncertainties are introduced in Section 4.3. For this purpose, we consider three scenarios ξ_1, ξ_2, and ξ_3. Scenario ξ_2 corresponds to values of the problem without uncertainties discussed in Section 7.3. In scenario ξ_1, we assume all uncertain values to take their lower bounds within the uncertainty range and in scenario ξ_3 their upper bounds, respectively. However, any other scenario could be chosen.

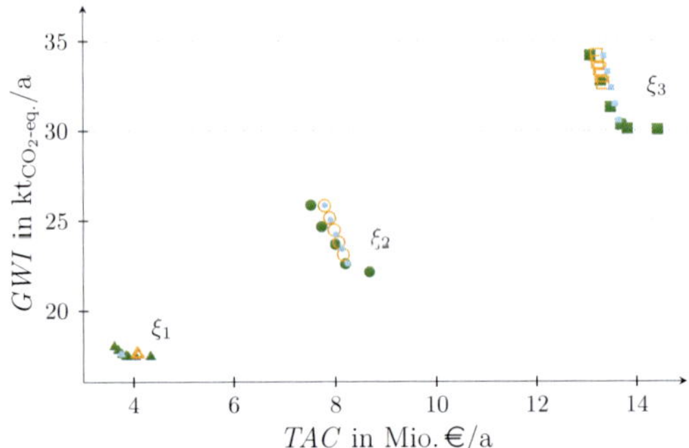

Figure D.3: Triangles, circles, and squares represent scenario ξ_1, ξ_2, and ξ_3, respectively; dark green filled marks: ideal Pareto front generated for each scenario separately, orange unfilled marks: robust flex-hand Pareto front in each scenario; small light blue marks: flex-hand Pareto front for separately considered scenarios; Pareto fronts are presented without normalization

To evaluate the robust flex-hand design, we compare the robust flex-hand Pareto fronts in all three scenarios to the flex-hand Pareto fronts generated for each scenario separately. Fig. D.3 shows that the flex-hand Pareto fronts generated for each scenario separately do not coincide with the robust flex-hand Pareto fronts. In scenario ξ_3, the robust flex-hand design leads to smaller total annualized costs than the flex-hand design computed for scenario ξ_3 but to a higher global warming impact. In total, the robust flex-hand Pareto front is less "streched out" than for the nominal case (Section 7.3) leading to an optimal distance $\bar{\varepsilon}^*$ of 0.625. The reduced adaptability to the ideal Pareto fronts is due to the fact that the robust flex-hand Pareto fronts need to approximate three ideal Pareto fronts simultaneously, instead of just one Pareto front. Thus, a good performance of a flex-hand design in one scenario might lead to a poor performance in another scenario if uncertainties are not regarded during design (Fig. D.4). In contrast, the robust flex-hand design is a compromise solution performing well in all three scenarios simultaneously.

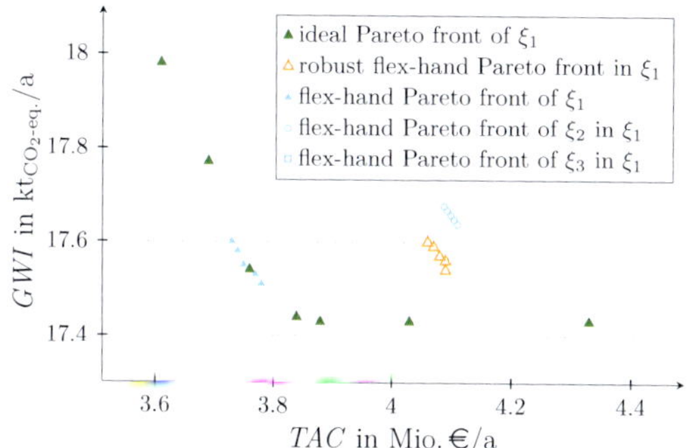

Figure D.4: All Pareto fronts for scenario ξ_1; dark green filled triangles (▲): ideal Pareto front of scenario ξ_1; orange unfilled triangles (△): robust flex-hand Pareto front in scenario ξ_1; small light blue triangles (▲): flex-hand Pareto front of scenario ξ_1; small light blue unfilled circles and squares (○ and □): flex-hand Pareto front in scenario ξ_1 based on flex-hand design computed for scenario ξ_2 and ξ_3, respectively; Pareto fronts are presented without normalization

In Fig. D.4, we take a closer look on the computed Pareto fronts in scenario ξ_1. In scenario ξ_1, the robust flex-hand design (△) clearly performs better than the flex-hand design selected for scenario ξ_2 (○) and the flex-hand design selected for scenario ξ_3 (□). In scenario ξ_1, only the flex-hand Pareto front of scenario ξ_1 (▲) approximates the ideal Pareto front (▲) better than the robust flex-hand Pareto front (△). However, the flex-hand design of scenario ξ_1 is infeasible for scenario ξ_2 and ξ_3. In contrast, the robust flex-hand design is feasible and performs well for all scenarios.

Having a closer look at the design (Fig. D.5), we observe that the total installed thermal power of the three flex-hand designs increases from scenario ξ_1 to ξ_3. This is due to the fact that values of uncertain input parameters increase as well. With increasing demands and also increasing specific global warming impact of the electricity grid, larger combined heat and power engines and boilers are installed combined with a higher installed thermal power of absorption chillers and smaller compression chillers. The robust flex-hand design does not differ remarkably from the three flex-hand designs. Thus, the scenario-based flex-hand approach is necessary to identify the excellent compromise given by the robust flex-hand design.

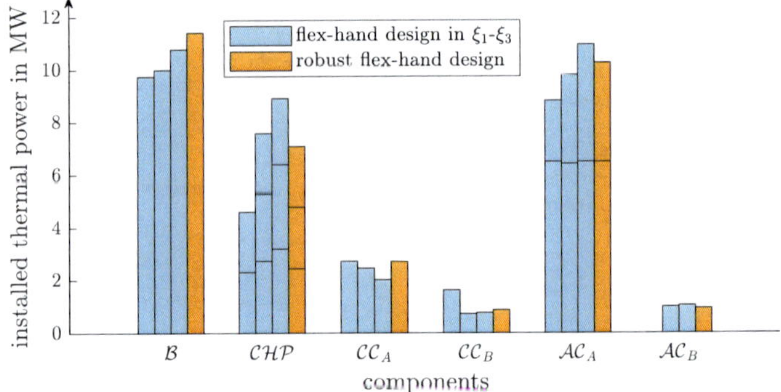

Figure D.5: In light blue: flex-hand designs generated for each scenario separately (from left to right for each component type: scenario ξ_1, ξ_2, and ξ_3); in orange: robust flex-hand design; $\mathcal{B}$ boiler, $\mathcal{CHP}$ combined heat and power engine, $\mathcal{CC}_A$ and $\mathcal{CC}_B$ compression chillers, and $\mathcal{AC}_A$ and $\mathcal{AC}_B$ absorption chillers installed on Site A and Site B, respectively

Appendix E

The be-rebust design

This appendix provides supplemental material for Chapter 8.

E.1 Problem formulation to select the be-rebust design of DESS

In this chapter, the problem formulation of the be-rebust framework of DESS is provided.

E.1.1 Calculation of the ideal reliable and robust Pareto front

The following problem formulation integrates $(n-1^{max})$-reliability into RoMOand, thus, enables to calculate the ideal reliable and robust Pareto front.

$$\min \begin{pmatrix} \alpha^{TAC} \\ \alpha^{GWI} \end{pmatrix}$$

s. t. objective functions

$$\sum_{t \in \mathcal{T}} \Big[\Delta\tau_t \Big(\widehat{p}^{gas}(1+\delta^{pg}) \cdot \sum_{k \in \mathcal{K}^h} \dot{U}_{kt} + \widehat{p}^{el,buy}(1+pe) \cdot U_t^{el,buy} - \widehat{p}^{el,sell}(1+pe) \cdot \dot{V}_t^{el,sell} \Big) \Big] + \sum_{k \in \mathcal{K}} \left(\frac{1}{PVF} + p_k^m \right) \cdot \underbrace{\sum_{h \in [H]} \Big[\gamma_{kh} \cdot \kappa_{kh} + m_{kh} \cdot \Big(\dot{V}_{kh}^N - \gamma_{kh} \dot{V}_{kh}^{N,lb} \Big) \Big]}_{=:I_k} \leq \alpha^{TAC} \quad \forall pe \in \{-\delta^{pe}, \delta^{pe}\}$$

$$\sum_{t \in \mathcal{T}} \Delta\tau_t \Big[\sum_{k \in \mathcal{K}^h} \dot{U}_{kt} \cdot GWI^{gas} + \Big(\dot{U}_t^{el,buy} - \dot{V}_t^{el,sell} \Big) \cdot \Big(\widehat{GWI}^{el} + ge \Big) \Big] \leq \alpha^{GWI} \quad \forall ge \in \{-\underline{\delta}^{ge}, \overline{\delta}^{ge}\}$$

robust, nominal, and lower bound energy balances

$$\sum_{k \in \mathcal{K}^h} \dot{V}_{kt} - \sum_{k \in \mathcal{AC}} \dot{U}_{kt} \geq \widehat{\dot{E}}_t^h + \delta_t^{\dot{E}h} \qquad \forall t \in \mathcal{T}$$

$$\sum_{k \in \mathcal{K}^c} \dot{V}_{kt} \geq \widehat{\dot{E}}_t^c + \delta_t^{\dot{E}c} \qquad \forall t \in \mathcal{T}$$

$$\sum_{k \in \mathcal{CHP}} \dot{V}_{kt}^{el} - \sum_{k \in \mathcal{CC}} \dot{U}_{kt}^{el} + \dot{U}_t^{el,buy} - \dot{V}_t^{el,sell} \geq \widehat{\dot{E}}_t^{el} + \delta_t^{\dot{E}e} \qquad \forall t \in \mathcal{T}$$

$$\sum_{k \in \mathcal{K}^h} \widehat{\dot{V}}_{kt} - \sum_{k \in \mathcal{AC}} \widehat{\dot{U}}_{kt} = \widehat{\dot{E}}_t^h \qquad \forall t \in \mathcal{T}$$

$$\sum_{k \in \mathcal{K}^c} \widehat{\dot{V}}_{kt} = \widehat{\dot{E}}_t^c \qquad \forall t \in \mathcal{T}$$

$$\sum_{k \in \mathcal{CHP}} \widehat{\dot{V}}_{kt}^{el} - \sum_{k \in \mathcal{CC}} \widehat{\dot{U}}_{kt}^{el} + \widehat{\dot{U}}_t^{el,buy} - \widehat{\dot{V}}_t^{el,sell} = \widehat{\dot{E}}_t^{el} \qquad \forall t \in \mathcal{T}$$

$$\sum_{k \in \mathcal{K}^h} \underline{\dot{V}}_{kt} - \sum_{k \in \mathcal{AC}} \underline{\dot{U}}_{kt} = \widehat{\dot{E}}_t^h - \delta_t^{\dot{E}h} \qquad \forall t \in \mathcal{T}$$

$$\sum_{k \in \mathcal{K}^c} \underline{\dot{V}}_{kt} = \widehat{\dot{E}}_t^c - \delta_t^{\dot{E}c} \qquad \forall t \in \mathcal{T}$$

$$\sum_{k\in\mathcal{CHP}} \underline{\dot{V}}_{kt}^{el} - \sum_{k\in\mathcal{CC}} \underline{\dot{U}}_{kt}^{el} + \underline{\dot{U}}_{t}^{el,buy} - \underline{\dot{V}}_{t}^{el,sell} = \hat{\dot{E}}_{t}^{el} - \delta_t^{\dot{E}e} \qquad \forall t \in \mathcal{T}$$

reliability constraints with integrated robustness

$$\sum_{k\in\mathcal{K}^c} \dot{V}_k^N - \dot{V}_{\mathcal{K}^c}^{max} \geq \hat{\dot{E}}_t^c + \delta_t^{\dot{E}c} \qquad \forall t \in \mathcal{T}$$

$$\sum_{k\in\mathcal{K}^h} \dot{V}_k^N - \dot{V}_{\mathcal{K}^h}^{max} \geq \hat{\dot{E}}_t^h + \delta_t^{\dot{E}h} + \sum_{k\in\mathcal{AC}} \frac{\dot{V}_{kt}}{\eta_k} \qquad \forall t \in \mathcal{T}$$

$$\sum_{k\in\mathcal{K}^h} \dot{V}_k^N \geq \hat{\dot{E}}_t^h + \delta_t^{\dot{E}h} + \sum_{k\in\mathcal{AC}} \frac{\dot{V}_{kt}}{\eta_k} + \dot{E}_{\mathcal{AC}}^h \qquad \forall t \in \mathcal{T}$$

$$\sum_{k\in\mathcal{K}^h} \dot{V}_k^N - \dot{V}_{\mathcal{K}^h}^{max} \geq \hat{\dot{E}}_t^h + \sum_{k\in\mathcal{AC}} \frac{\hat{\dot{V}}_{kt}}{\eta_k} \qquad \forall t \in \mathcal{T}$$

$$\sum_{k\in\mathcal{K}^h} \dot{V}_k^N \geq \hat{\dot{E}}_t^h + \sum_{k\in\mathcal{AC}} \frac{\hat{\dot{V}}_{kt}}{\eta_k} + \dot{E}_{\mathcal{AC}}^h \qquad \forall t \in \mathcal{T}$$

$$\sum_{k\in\mathcal{K}^h} \dot{V}_k^N - \dot{V}_{\mathcal{K}^h}^{max} \geq \hat{\dot{E}}_t^h - \delta_t^{\dot{E}h} + \sum_{k\in\mathcal{AC}} \frac{\underline{\dot{V}}_{kt}}{\eta_k} \qquad \forall t \in \mathcal{T}$$

$$\sum_{k\in\mathcal{K}^h} \dot{V}_k^N \geq \hat{\dot{E}}_t^h - \delta_t^{\dot{E}h} + \sum_{k\in\mathcal{AC}} \frac{\underline{\dot{V}}_{kt}}{\eta_k} + \dot{E}_{\mathcal{AC}}^h \qquad \forall t \in \mathcal{T}$$

$$\dot{E}_{\mathcal{AC}}^h \geq x \cdot \frac{\dot{V}_{\mathcal{CC}}^{max}}{\eta_{\mathcal{AC}}^{min}}$$

$$\eta_{\mathcal{AC}}^{min} \leq \eta_k \qquad \forall k \in \mathcal{AC}$$

$$\dot{V}_{\mathcal{CC}}^{max} \geq \dot{V}_k^N \qquad \forall k \in \mathcal{CC}$$

$$\dot{E}_{\mathcal{AC}}^h \geq y \cdot \sum_{k\in\mathcal{AC}} \frac{\dot{V}_k^N}{\eta_k}$$

$$x + y = 1$$

linearization of investment costs

$$\sum_{h\in[H]} \gamma_{kh} \leq 1 \qquad \forall k \in \mathcal{K}$$

$$\gamma_{kh} \cdot \dot{V}_{kh}^{N,lb} \leq \dot{V}_{kh}^N \leq \gamma_{kh} \cdot \dot{V}_{kh+1}^{N,lb} \qquad \forall k \in \mathcal{K}, h \in [H]$$

further constraints and variables

$$\mu^{min} \cdot \sum_{h\in[H]} \dot{V}_{kh}^N \leq \dot{V}_{kt} \leq \sum_{h\in[H]} \dot{V}_{kh}^N \qquad \forall k \in \mathcal{K}, t \in \mathcal{T}$$

$$\mu^{min} \cdot \sum_{h\in[H]} \dot{V}_{kh}^N \leq \hat{\dot{V}}_{kt} \leq \sum_{h\in[H]} \dot{V}_{kh}^N \qquad \forall k \in \mathcal{K}, t \in \mathcal{T}$$

$$
\begin{aligned}
&\mu^{min} \cdot \sum_{h \in [H]} \dot{V}^N_{kh} \leq \underline{\dot{V}}_{kt} \leq \sum_{h \in [H]} \dot{V}^N_{kh} && \forall k \in \mathcal{K}, t \in \mathcal{T} \\
&\dot{V}_{kt} = \eta_k \cdot \dot{U}_{kt} && \forall k \in \mathcal{K}, t \in \mathcal{T} \\
&\widehat{\dot{V}}_{kt} = \eta_k \cdot \widehat{\dot{U}}_{kt} && \forall k \in \mathcal{K}, t \in \mathcal{T} \\
&\underline{\dot{V}}_{kt} = \eta_k \cdot \underline{\dot{U}}_{kt} && \forall k \in \mathcal{K}, t \in \mathcal{T} \\
&\dot{V}^{el}_{kt} = \eta^{tot}_k \cdot \dot{U}_{kt} - \dot{V}_{kt} && \forall k \in \mathcal{CHP}, t \in \mathcal{T} \\
&\widehat{\dot{V}}^{el}_{kt} = \eta^{tot}_k \cdot \widehat{\dot{U}}_{kt} - \widehat{\dot{V}}_{kt} && \forall k \in \mathcal{CHP}, t \in \mathcal{T} \\
&\underline{\dot{V}}^{el}_{kt} = \eta^{tot}_k \cdot \underline{\dot{U}}_{kt} - \underline{\dot{V}}_{kt} && \forall k \in \mathcal{CHP}, t \in \mathcal{T} \\
&\dot{U}^{el,buy}, \dot{V}^{el,sell}, \dot{V}^{el}, \dot{U}, \dot{V}, \widehat{\dot{U}}, \widehat{\dot{V}}, \underline{\dot{U}}, \underline{\dot{V}} \in \mathbb{R}^{|\mathcal{K}| \times T}_+ \\
&x, y \in \{0,1\}, \gamma \in \{0,1\}^{|\mathcal{K}| \times H}, \dot{V}^N \in \mathbb{R}^{|\mathcal{K}| \times H}_+ ,
\end{aligned}
$$

with cooling components $\mathcal{K}^c = \mathcal{AC} \cup \mathcal{CC}$ and heating components $\mathcal{K}^h = \mathcal{B} \cup \mathcal{CHP}$. For further declaration of symbols, see Appendix A.2 or the Nomenclature.

E.1.2 Selection of the be-rebust design

In this section, the problem formulation for the be-rebust framework is given. The adaption of the flex-hand approach for enabling application to RoMO with included $(n-1^{max})$-reliability is provided in the following. All equations which do not depend on j remain identical to the problem formulation of RoMO and are only listed in the following for completeness.

$$
\begin{aligned}
&\min \ \overline{\varepsilon} \\
&\text{s.t. } \overline{\alpha^{TAC}}(j) - \left(\overline{TAC^{rr}}\right)^*(j) \leq \overline{\varepsilon} && \forall j \in [N] \\
&\quad\ \ \overline{\alpha^{GWI}}(j) - \left(\overline{GWI^{rr}}\right)^*(j) \leq \overline{\varepsilon} && \forall j \in [N]
\end{aligned}
$$

objective functions

$$\sum_{t\in\mathcal{T}}\Big[\Delta\tau_t\Big(\widehat{p}^{gas}(1+\delta^{pg})\cdot\sum_{k\in\mathcal{K}^h}\dot{U}_{kt}(j) + \widehat{p}^{el,buy}(1+pe)\cdot\dot{U}_t^{el,buy}(j) - \widehat{p}^{el,sell}(1+pe)\cdot\dot{V}_t^{el,sell}(j)\Big)\Big] + \sum_{k\in\mathcal{K}}\left(\frac{1}{PVF}+p_k^m\right)\cdot\underbrace{\sum_{h\in[H]}\Big[\gamma_{kh}\cdot\kappa_{kh}+m_{kh}\cdot\Big(V_{kh}^N-\gamma_{kh}\dot{V}_{kh}^{N,lb}\Big)\Big]}_{=:I_k}\le\alpha^{TAC}(j) \qquad \forall pe\in\{-\delta^{pe},\delta^{pe}\}\ \ \forall j\in[N]$$

$$\sum_{t\in\mathcal{T}}\Delta\tau_t\Bigg[\sum_{k\in\mathcal{K}^h}\dot{U}_{kt}(j)\cdot GWI^{gas} + \Big(\dot{U}_t^{el,buy}(j)-\dot{V}_t^{el,sell}(j)\Big)\cdot\Big(\widehat{GWI}^{el}+ge\Big)\Bigg]\le\alpha^{GWI}(j) \qquad \forall ge\in\{-\underline{\delta}^{ge},\overline{\delta}^{ge}\}\ \ \forall j\in[N]$$

robust, nominal, and lower bound energy balances

$$\sum_{k\in\mathcal{K}^h}\dot{V}_{kt}(j)-\sum_{k\in\mathcal{AC}}\dot{U}_{kt}(j)\ge\widehat{\dot{E}}_t^h+\delta_t^{\dot{E}h} \qquad \forall t\in\mathcal{T},j\in[N]$$

$$\sum_{k\in\mathcal{K}^c}\dot{V}_{kt}(j)\ge\widehat{\dot{E}}_t^c+\delta_t^{\dot{E}c} \qquad \forall t\in\mathcal{T},j\in[N]$$

$$\sum_{k\in\mathcal{CHP}}\dot{V}_{kt}^{el}(j)-\sum_{k\in\mathcal{CC}}\dot{U}_{kt}^{el}(j)+\dot{U}_t^{el,buy}(j)-\dot{V}_t^{el,sell}(j)\ge\widehat{\dot{E}}_t^{el}+\delta_t^{\dot{E}e} \qquad \forall t\in\mathcal{T},j\in[N]$$

$$\sum_{k\in\mathcal{K}^h}\widehat{\dot{V}}_{kt}-\sum_{k\in\mathcal{AC}}\widehat{\dot{U}}_{kt}=\widehat{\dot{E}}_t^h \qquad \forall t\in\mathcal{T}$$

$$\sum_{k\in\mathcal{K}^c}\widehat{\dot{V}}_{kt}=\widehat{\dot{E}}_t^c \qquad \forall t\in\mathcal{T}$$

$$\sum_{k\in\mathcal{CHP}}\widehat{\dot{V}}_{kt}^{el}-\sum_{k\in\mathcal{CC}}\widehat{\dot{U}}_{kt}^{el}+\widehat{\dot{U}}_t^{el,buy}-\widehat{\dot{V}}_t^{el,sell}=\widehat{\dot{E}}_t^{el} \qquad \forall t\in\mathcal{T}$$

$$\sum_{k\in\mathcal{K}^h}\underline{\dot{V}}_{kt}-\sum_{k\in\mathcal{AC}}\underline{\dot{U}}_{kt}=\widehat{\dot{E}}_t^h-\delta_t^{\dot{E}h} \qquad \forall t\in\mathcal{T}$$

$$\sum_{k\in\mathcal{K}^c}\underline{\dot{V}}_{kt}=\widehat{\dot{E}}_t^c-\delta_t^{\dot{E}c} \qquad \forall t\in\mathcal{T}$$

$$\sum_{k\in\mathcal{CHP}}\underline{\dot{V}}_{kt}^{el}-\sum_{k\in\mathcal{CC}}\underline{\dot{U}}_{kt}^{el}+\underline{\dot{U}}_t^{el,buy}-\underline{\dot{V}}_t^{el,sell}=\widehat{\dot{E}}_t^{el}-\delta_t^{\dot{E}e} \qquad \forall t\in\mathcal{T}$$

reliability constraints with integrated robustness

$$\sum_{k\in\mathcal{K}^c} \dot{V}_k^N - \dot{V}_{\mathcal{K}^c}^{max} \geq \hat{\dot{E}}_t^c + \delta_t^{\dot{E}c} \qquad \forall t\in\mathcal{T}$$

$$\sum_{k\in\mathcal{K}^h} \dot{V}_k^N - \dot{V}_{\mathcal{K}^h}^{max} \geq \hat{\dot{E}}_t^h + \delta_t^{\dot{E}h} + \sum_{k\in\mathcal{AC}} \frac{\dot{V}_{kt}(j)}{\eta_k} \qquad \forall t\in\mathcal{T}, j\in[N]$$

$$\sum_{k\in\mathcal{K}^h} \dot{V}_k^N \geq \hat{\dot{E}}_t^h + \delta_t^{\dot{E}h} + \sum_{k\in\mathcal{AC}} \frac{\dot{V}_{kt}(j)}{\eta_k} + \dot{E}_{\mathcal{AC}}^h \qquad \forall t\in\mathcal{T}, j\in[N]$$

$$\sum_{k\in\mathcal{K}^h} \dot{V}_k^N - \dot{V}_{\mathcal{K}^h}^{max} \geq \hat{\dot{E}}_t^h + \sum_{k\in\mathcal{AC}} \frac{\hat{\dot{V}}_{kt}}{\eta_k} \qquad \forall t\in\mathcal{T}$$

$$\sum_{k\in\mathcal{K}^h} \dot{V}_k^N \geq \hat{\dot{E}}_t^h + \sum_{k\in\mathcal{AC}} \frac{\hat{\dot{V}}_{kt}}{\eta_k} + \dot{E}_{\mathcal{AC}}^h \qquad \forall t\in\mathcal{T}$$

$$\sum_{k\in\mathcal{K}^h} \dot{V}_k^N - \dot{V}_{\mathcal{K}^h}^{max} \geq \hat{\dot{E}}_t^h - \delta_t^{\dot{E}h} + \sum_{k\in\mathcal{AC}} \frac{\underline{\dot{V}}_{kt}}{\eta_k} \qquad \forall t\in\mathcal{T}$$

$$\sum_{k\in\mathcal{K}^h} \dot{V}_k^N \geq \hat{\dot{E}}_t^h - \delta_t^{\dot{E}h} + \sum_{k\in\mathcal{AC}} \frac{\underline{\dot{V}}_{kt}}{\eta_k} + \dot{E}_{\mathcal{AC}}^h \qquad \forall t\in\mathcal{T}$$

$$\dot{V}_{\mathcal{K}^{h/c}}^{max} \geq \dot{V}_k^N \qquad \forall k\in\mathcal{K}^{h/c}$$

$$\dot{E}_{\mathcal{AC}}^h \geq x \cdot \frac{\dot{V}_{\mathcal{CC}}^{max}}{\eta_{\mathcal{AC}}^{min}}$$

$$\eta_{\mathcal{AC}}^{min} \leq \eta_k \qquad \forall k\in\mathcal{AC}$$

$$\dot{V}_{\mathcal{CC}}^{max} \geq \dot{V}_k^N \qquad \forall k\in\mathcal{CC}$$

$$\dot{E}_{\mathcal{AC}}^h \geq y \cdot \sum_{k\in\mathcal{AC}} \frac{\dot{V}_k^N}{\eta_k}$$

$$x + y = 1$$

linearization of investment costs

$$\sum_{h\in[H]} \gamma_{kh} \leq 1 \qquad \forall k\in\mathcal{K}$$

$$\gamma_{kh} \cdot \dot{V}_{kh}^{N,lb} \leq \dot{V}_{kh}^N \leq \gamma_{kh} \cdot \dot{V}_{kh+1}^{N,lb} \qquad \forall k\in\mathcal{K}, h\in[H]$$

further constraints and variables

$$\mu^{min} \cdot \sum_{h\in[H]} \dot{V}_{kh}^N \leq \dot{V}_{kt}(j) \leq \sum_{h\in[H]} \dot{V}_{kh}^N \qquad \forall k\in\mathcal{K}, t\in\mathcal{T}, j\in[N]$$

$$\mu^{min} \cdot \sum_{h\in[H]} \dot{V}^N_{kh} \leq \widehat{\dot{V}}_{kt} \leq \sum_{h\in[H]} \dot{V}^N_{kh} \qquad \forall k \in \mathcal{K}, t \in \mathcal{T}$$

$$\mu^{min} \cdot \sum_{h\in[H]} \dot{V}^N_{kh} \leq \underline{\dot{V}}_{kt} \leq \sum_{h\in[H]} \dot{V}^N_{kh} \qquad \forall k \in \mathcal{K}, t \in \mathcal{T}$$

$$\dot{V}_{kt}(j) = \eta_k \cdot \dot{U}_{kt}(j) \qquad \forall k \in \mathcal{K}, t \in \mathcal{T}, j \in [N]$$

$$\widehat{\dot{V}}_{kt} = \eta_k \cdot \widehat{\dot{U}}_{kt} \qquad \forall k \in \mathcal{K}, t \in \mathcal{T}$$

$$\underline{\dot{V}}_{kt} = \eta_k \cdot \underline{\dot{U}}_{kt} \qquad \forall k \in \mathcal{K}, t \in \mathcal{T}$$

$$\dot{V}^{el}_{kt}(j) = \eta^{tot}_k \cdot \dot{U}_{kt}(j) - \dot{V}_{kt}(j) \qquad \forall k \in \mathcal{CHP}, t \in \mathcal{T}, j \in [N]$$

$$\widehat{\dot{V}}^{el}_{kt} = \eta^{tot}_k \cdot \widehat{\dot{U}}_{kt} - \widehat{\dot{V}}_{kt} \qquad \forall k \in \mathcal{CHP}, t \in \mathcal{T}$$

$$\underline{\dot{V}}^{el}_{kt} = \eta^{tot}_k \cdot \underline{\dot{U}}_{kt} - \underline{\dot{V}}_{kt} \qquad \forall k \in \mathcal{CHP}, t \in \mathcal{T}$$

$$\dot{U}^{el,buy}(j), \dot{V}^{el,sell}(j), \dot{V}^{el}(j), \dot{U}(j), \dot{V}(j) \in \mathbb{R}^{|\mathcal{K}|\times T}_+ \qquad \forall j \in [N]$$

$$\widehat{\dot{U}}, \widehat{\dot{V}}, \underline{\dot{U}}, \underline{\dot{V}} \in \mathbb{R}^{|\mathcal{K}|\times T}_+$$

$$x, y \in \{0,1\}, \gamma \in \{0,1\}^{|\mathcal{K}|\times H}, \dot{V}^N \in \mathbb{R}^{|\mathcal{K}|\times H}_+$$

$$\overline{\varepsilon} \in \mathbb{R}_+ \, ,$$

with cooling components $\mathcal{K}^c = \mathcal{AC} \cup \mathcal{CC}$ and heating components $\mathcal{K}^h = \mathcal{B} \cup \mathcal{CHP}$. Bars above the total annualized costs TAC and the global warming impact GWI in the optimization problem denote the normalization of the objective values. Objective values on the normalized ideal reliable and robust Pareto fronts based on the integration of $(n - 1^{max})$-reliability into RoMO (Section 8.1.2) are denoted by $\left(\left(\overline{TAC}^{rr}\right)^*(j), \left(\overline{GWI}^{rr}\right)^*(j) \right)$ for each point $j \in [N]$. For further declaration of symbols see Appendix A.2 or the Nomenclature.

E.2 Further results of the case study

In this section, further results of the case study of Chapter 8 are presented.

E.2.1 The be-rebust design

Even though there is only a small trade-off, the installed thermal power of combined heat and power engines $\mathcal{CHP}$ increases with slightly decreasing global warming impact GWI^{rr}, while the installed thermal power of boilers $\mathcal{B}$ decreases (Fig. E.1). However, the be-rebust design does not show remarkable differences compared to the other design options.

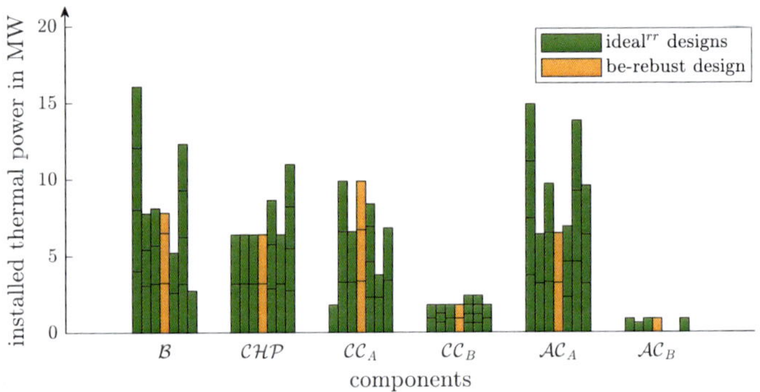

Figure E.1: In dark green: designs of ideal^{rr} Pareto front; in orange: the be-rebust design; from left to right for each component type, the designs are ordered by decreasing minimal global warming impact; $\mathcal{B}$ boiler, $\mathcal{CHP}$ combined heat and power engine, $\mathcal{CC}_A$ and $\mathcal{CC}_B$ compression chillers, and $\mathcal{AC}_A$ and $\mathcal{AC}_B$ absorption chillers installed on Site A and Site B, respectively; rr highlights that solutions are reliable and robust, be-rebust designs shortens the expression be-rebust design

As performed in Section 8.2.2 for the uncertainty-scaling factor $\omega = 0.66$), we re-optimize operation of the be-rebust design (for $\omega = 1$) without taking uncertainties of supply and input parameters into account (Fig. E.2).

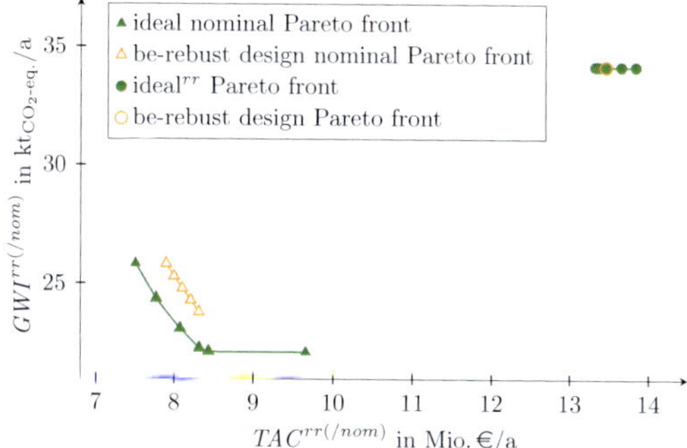

Figure E.2: Comparison of nominal Pareto front (dark green triangles) and the nominal Pareto front based on the be-rebust design (orange unfilled triangles); the ideal^{rr} Pareto front and the be-rebust design Pareto front are presented only for comparison, (see also Fig. E.1); rr highlights that solutions are reliable and robust; Pareto fronts are presented without normalization; lines are included to guide reader's eyes

The comparison of the Pareto fronts shows that the be-rebust design performs well in the everyday business. The distance of the minimal nominal total annualized costs $TAC^{rr/nom}$ to the minimal nominal total annualized costs TAC^{*} is only 5.2 %.

E.2.2 Sensitivity analysis of the be-rebust design

A sensitivity analysis of the be-rebust design on the uncertainty of input parameters is performed in Section 8.2.2. The designs of ideal reliable and robust Pareto front as well as the flex-hand reliable and robust design with respect to uncertainty-scaling factor $\omega = 0.33$ are presented in Fig. E.3.

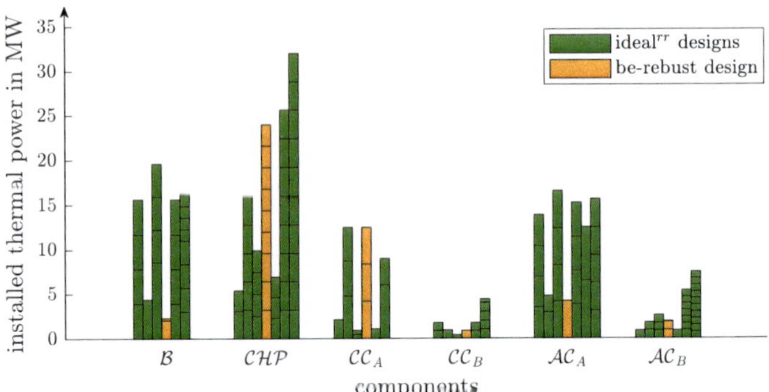

Figure E.3: In dark green: designs of ideal^{rr} Pareto front; in orange: the be-rebust design with respect to uncertainty-scaling factor $\omega = 0.33$; from left to right for each component type, the designs are ordered by decreasing minimal global warming impact; $\mathcal{B}$ boiler, $\mathcal{CHP}$ combined heat and power engine, $\mathcal{CC}_A$ and $\mathcal{CC}_B$ compression chillers, and $\mathcal{AC}_A$ and $\mathcal{AC}_B$ absorption chillers installed on Site A and Site B, respectively; rr highlights that solutions are reliable and robust, be-rebust designs shortens the expression be-rebust design

Appendix F

List of Publications

Journal Papers

Baumgärtner, N., Shu, D. Y., Bahl, B., Hennen, M., Hollermann, D. E., and Bardow, A. (2020). DeLoop: Decomposition based long-term operational optimization of energy systems with time-coupling constraints. *Energy*, 117272

Sass, S., Faulwasser, T., Hollermann, D. E., Kappatou, C. D., Sauer, D., Schütz, T., Shu, D. Y., Bardow, A., Gröll, L., Hagenmeyer, V., Müller, D., and Mitsos, A. (2020). Model compendium, data, and optimization benchmarks for sector-coupled energy systems. *Computers & Chemical Engineering*, 106760

Hollermann, D. E., Goerigk, M., Hoffrogge, D. F., Hennen, M. R., and Bardow, A. (2019b). Flexible here-and-now decisions for two-stage multi-objective optimization: Method and application to energy system design selection. *arXiv e-prints*, arXiv:1906.08621

Hollermann, D. E., Hoffrogge, D. F., Mayer, F., Hennen, M. R., and Bardow, A. (2019a). Optimal $(n-1)$-reliable design of distributed energy supply systems. *Computers & Chemical Engineering*, 121:317–326

Bahl, B., Lützow, J., Shu, D., Hollermann, D. E., Lampe, M., and Bardow, A. (2018). Rigorous synthesis of energy systems by decomposition via time-series aggregation. *Computers & Chemical Engineering*, 112:70–81

Majewski, D. E., Wirtz, M., Lampe, M., and Bardow, A. (2017a). Robust multi-objective optimization for sustainable design of distributed energy supply systems. *Computers & Chemical Engineering*, 102:26–39

Majewski, D. E., Lampe, M., Voll, P., and Bardow, A. (2017b). TRusT: A Two-stage Robustness Trade-off approach for the design of decentralized energy supply systems. *Energy*, 118:590–599

Bardow, A., Majewski, D. E., Madlener, R., Monti, A., Müller, D., and Streblow, R. (2015). Effizienz – Die unvergessene nachhaltige Energiequelle: Energieeinsparungen und ihre Rebound-Effekte. *RWTH Themen*, 1:22–26

Conference Contributions

Baumgärtner, N. J., Temme, F., Bahl, B., Hennen, M. R., Hollermann, D. E., Bardow, A. (2019). RiSES4 Rigorous Synthesis of Energy Supply Systems with Seasonal Storage by relaxation and time-series aggregation to typical periods. *ECOS 2019 – The 32nd International Conference on Efficiency, Cost, Optimization, Simulation and Environmental Impact of Energy Systems.* Oral presentation. Wroclaw, Poland.

Hollermann, D. E., Hoffrogge, D. F., Hennen, M. R., Bardow, A. (2018). Ensuring $(n-1)$-reliability in the optimal design of distributed energy supply systems. *ESCAPE28 – 28th European Symposium on Computer Aided Process Engineering, 307–312.* Oral presentation. Graz, Switzerland.

Majewski, D. E., Jensen, M., Hennen, M. R., and Bardow, A. (2017). Reliable optimal operation of distributed energy supply systems. *COST TD1207.* Oral presentation. Modena, Italy.

Majewski, D. E., Wirtz, M., Lampe, M., and Bardow, A. (2016). TRusT: A Two-stage Robustness Trade-off approach for future energy systems. *PhD Workshop of the joint initiative "Energy System 2050".* Oral presentation. Hamburg, Germany.

Bahl, B., Lützow, J., Majewski, D. E., Lampe, M., Hennen, M. R., Bardow, A. (2017). Rigorous synthesis of energy supply systems by time-series aggregation. *ESCAPE27 – 27th European Symposium on Computer Aided Process Engineering, 2413–2418.* Oral presentation. Barcelona, Spain.

Hennen, M. R., Bahl, B., Hollermann, D. E., Holters, L., Bardow, A. (2017). Optimal design of distributed energy supply systems. *Sustainable Process*

Integration Laboratory Scientific Conference, 2413–2418. Poster presentation. Brno, Czech Republic.

Majewski, D. E., Wirtz, M., Lampe, M., and Bardow, A. (2016). Two-stage Robustness Trade-off: TRusT in the design of energy systems. *OR 2016.* Oral presentation. Hamburg, Germany.

Majewski, D. E., Voll, P., Bardow, A. (2015). Strictly robust optimal design of decentralized energy systems. *ECOS 2016 – 28th International Conference on Efficiency, Cost, Optimization, Simulation and Environmental Impact of Energy Systems*, 2:751–762. Oral presentation. Pau, France.

Majewski, D. E., Voll, P., Bardow, A. (2015). Synthesis of energy supply systems by strictly robust optimization. *17th British-French-German Conference on Optimization.* Oral Presentation. London, United Kingdom.

Majewski, D. E., Voll, P., Bardow, A. (2015). Mixed-Integer Nonlinear Optimization Methods for Energy Systems Engineering. *Mixed-Integer Nonlinear Optimization Methods for Energy Systems Engineering.* Aachen, Germany.

Appendix G

Student theses supervised during this work

Mayer, Fabian (2018). *Two-stage robust and $(n-1)$-reliable optimization of distributed energy supply systems.* Master thesis, RWTH Aachen University.

Ihlemann, Maren (2017). *$(n-1)$-reliable structure optimization of decentralized energy systems considering start-up and shut-down processes.* Master thesis, RWTH Aachen University.

Kiesheyer, Bastian (2017). *Modellierung sowie ökonomische und ökologische Strukturoptimierung des Energieversorgungssystems eines Forschungscampus.* Bachelor thesis at BFT Planung GmbH, RWTH Aachen University.

Zachgo, Lars (2017). *Entwicklung und Evaluation von Methoden zur Zeitreihenadaption in der Betriebsoptimierung von industriellen Energiesystemen.* Master thesis at Athion GmbH, RWTH Aachen University.

Jensen, Malte (2016). *Entwicklung einer Methode zur systemverlässlichen Betriebsoptimierung von dezentralen Energieversorgungssystemen.* Master thesis, RWTH Aachen University.

Oltmanns, Jan-Gerd (2016). *Betriebsoptimierung eines industriellen Wärmeversorgungssystems mittels MILP.* Master thesis at perpendo Energie- und Verfahrenstechnik GmbH, RWTH Aachen University.

Hackbarth, Mirko (2016). *Entwicklung eines Optimierungstools für die Steuerung von dezentralen Energieerzeugungsanlagen im industriellen Umfeld.* Master thesis at Athion GmbH, RWTH Aachen University.

Feldhaus, Jerome (2016). *Entwicklung einer Zielfunktion und Bewertung von Methoden zur Positionierungsoptimierung in der Windparkplanung.* Master thesis at Fraunhofer IWES, RWTH Aachen University.

Wirtz, Marco (2015). *Robuste multikriterielle Optimierung von Energieversorgungssystemen.* Bachelor thesis, RWTH Aachen University.

Schneider, Georg Ferdinand (2014). *Optimization of Distributed Energy Supply Systems by Branch-and-Price.* Master thesis at the Chair of Operations Research, RWTH Aachen University.

Bibliography

Abdollahzadeh, H. and Atashgar, K. (2017). Optimal design of a multi-state system with uncertainty in supplier selection. *Computers & Chemical Engineering*, 105:411–434.

Abubaker, A., Baharum, A., and Alrefaei, M. (2014). Good solution for multi-objective optimization problem. *AIP Conference Proceedings*, 1605(1):1147–1152.

Aguilar, O., Kim, J.-K., Perry, S., and Smith, R. (2008). Availability and reliability considerations in the design and optimisation of flexible utility systems. *Chemical Engineering Science*, 63(14):3569–3584.

Aissi, H., Bazgan, C., and Vanderpooten, D. (2009). Min-max and min-max regret versions of combinatorial optimization problems: A survey. *European Journal of Operational Research*, 197(2):427–438.

Akbari, K., Nasiri, M. M., Jolai, F., and Ghaderi, S. F. (2014). Optimal investment and unit sizing of distributed energy systems under uncertainty: A robust optimization approach. *Energy and Buildings*, 85:275–286.

Alguacil, N., Carrión, M., and Arroyo, J. M. (2009). Transmission network expansion planning under deliberate outages. *International Journal of Electrical Power & Energy Systems*, 31(9):553–561.

Amusat, O. O., Shearing, P. R., and Fraga, E. S. (2017). On the design of complex energy systems: Accounting for renewables variability in systems sizing. *Computers & Chemical Engineering*, 103:103–115.

Andiappan, V. (2017). State-of-the-art review of mathematical optimisation approaches for synthesis of energy systems. *Process Integration and Optimization for Sustainability*, 1(3):165–188.

Andiappan, V. and Ng, D. K. (2016). Synthesis of tri-generation systems: Technology selection, sizing and redundancy allocation based on operational strategy. *Computers & Chemical Engineering*, 91:380–391.

Andiappan, V., Ng, D. K. S., and Tan, R. R. (2017). Design operability and retrofit analysis (DORA) framework for energy systems. *Energy*, 134:1038–1052.

Andiappan, V., Tan, R. R., Aviso, K. B., and Ng, D. K. (2015). Synthesis and optimisation of biomass-based tri-generation systems with reliability aspects. *Energy*, 89:803–818.

Antipova, E., Pozo, C., Guillén-Gosálbez, G., Boer, D., Cabeza, L., and Jiménez, L. (2015). On the use of filters to facilitate the post-optimal analysis of the Pareto solutions in multi-objective optimization. *Computers & Chemical Engineering*, 74:48–58.

Asprion, N., Blagov, S., Böttcher, R., Schwientek, J., Burger, J., von Harbou, E., and Bortz, M. (2017). Simulation and multi-criteria optimization under uncertain model parameters of a cumene process. *Chemie Ingenieur Technik*, 89(5):665–674.

Bahl, B., Lützow, J., Shu, D., Hollermann, D. E., Lampe, M., Hennen, M., and Bardow, A. (2018a). Rigorous synthesis of energy systems by decomposition via time-series aggregation. *Computers & Chemical Engineering*, 112:70–81.

Bahl, B., Söhler, T., Hennen, M., and Bardow, A. (2018b). Typical periods for two-stage synthesis by time-series aggregation with bounded error in objective function. *Frontiers in Energy Research*, 5:35.

Barlow, R. E., Proschan, F., and Hunter, L. C. (1965). *Mathematical Theory of Reliability*. Wiley, New York.

Baumann, H. and Tillman, A.-M. (2014). *The Hitch Hiker's Guide to LCA: An orientation in life cycle assessment methodology and application*. Studentlitteratur, Lund.

Baumgärtner, N., Bahl, B., Hennen, M., and Bardow, A. (2019a). RiSES3: Rigorous synthesis of energy supply and storage systems via time-series relaxation and aggregation. *Computers & Chemical Engineering*, 127:127–139.

Baumgärtner, N., Delorme, R., Hennen, M., and Bardow, A. (2019b). Design of low-carbon utility systems: Exploiting time-dependent grid emissions for climate-friendly demand-side management. *Applied Energy*, 247:755–765.

Baumgärtner, N., Shu, D., Bahl, B., Hennen, M., Hollermann, D. E., and Bardow, A. (2020). DeLoop: Decomposition-based long-term operational optimization of energy systems with time-coupling constraints. *Energy*, 198:117272.

Ben-Tal, A., Goryashko, A., Guslitzer, E., and Nemirovski, A. (2004). Adjustable robust solutions of uncertain linear programs. *Mathematical Programming*, 99(2):351–376.

Ben-Tal, A. and Nemirovski, A. (1999). Robust solutions of uncertain linear programs. *Operations Research Letters*, 25(1):1–13.

Ben-Tal, A. and Nemirovski, A. (2000). Robust solutions of linear programming problems contaminated with uncertain data. *Mathematical Programming*, 88(3):411–424.

Benders, J. (1962). Partitioning procedures for solving mixed-variables programming problems. *Numerische Mathematik*, 4(1):238–252.

Berndt, H., Hermann, M., Kreye, H. D., Reinisch, R., Scherer, U., and Vanzetta, J. (2007). *TransmissionCode 2007 – Network and System Rules of the German Transmission System Operators*. VDN e.V., Berlin.

Bertsimas, D., Litvinov, E., Sun, X. A., Zhao, J., and Zheng, T. (2013). Adaptive robust optimization for the security constrained unit commitment problem. *IEEE Transactions on Power Systems*, 28(1):52–63.

Bertsimas, D. and Sim, M. (2004). The price of robustness. *Operations Research*, 52(1):35–53.

Beume, N., Fonseca, C. M., Lopez-Ibanez, M., Paquete, L., and Vahrenhold, J. (2009). On the complexity of computing the hypervolume indicator. *IEEE Transactions on Evolutionary Computation*, 13(5):1075–1082.

Beykal, B., Boukouvala, F., Floudas, C. A., and Pistikopoulos, E. N. (2018). Optimal design of energy systems using constrained grey-box multi-objective optimization. *Computers & Chemical Engineering*, 116:488–502.

Birge, J. R. and Louveaux, F. (2011). *Introduction to Stochastic Programming*. Springer Series in Operations Research and Financial Engineering. Springer New York, 2nd edition.

Birnbaum, Z. W. (1968). On the importance of different components in a multi-component system. Technical report, Washington Univ Seattle Lab of Statistical Research.

Birol, F. (2018). We're getting closer to completing the energy transition.

Birol, F., Cozzi, L., Gould, T., Bromhead, A., and et al. (2015). World Energy Outlook 2015. International Energy Agency.

Bogenschneider, K. and Corbett, T. J. (2011). *Evidence-based policymaking: Insights from policy-minded researchers and research-minded policymakers*. Routledge.

Bokrantz, R. and Fredriksson, A. (2017). Necessary and sufficient conditions for Pareto efficiency in robust multiobjective optimization. *European Journal of Operational Research*, 262(2):682–692.

Bortz, M., Burger, J., Asprion, N., Blagov, S., Böttcher, R., Nowak, U., Scheithauer, A., Welke, R., Küfer, K.-H., and Hasse, H. (2014). Multi-criteria optimization in chemical process design and decision support by navigation on Pareto sets. *Computers & Chemical Engineering*, 60:354–363.

Bortz, M., Burger, J., von Harbou, E., Klein, M., Schwientek, J., Asprion, N., Böttcher, R., Küfer, K.-H., and Hasse, H. (2017). Efficient approach for calculating Pareto boundaries under uncertainties in chemical process design. *Industrial & Engineering Chemistry Research*, 56(44):12672–12681.

Botte, M. and Schöbel, A. (2019). Dominance for multi-objective robust optimization concepts. *European Journal of Operational Research*, 273(2):430–440.

Branke, J., Deb, K., Dierolf, H., and Osswald, M. (2004). Finding knees in multi-objective optimization. In *Parallel Problem Solving from Nature – PPSN VIII*, pages 722–731. Springer Berlin Heidelberg.

Broverman, S. A. (2010). *Mathematics of Investment and Credit*. ACTEX Publications, Inc., New Hartford, 5th edition.

Bruno, J., Fernandez, F., Castells, F., and Grossmann, I. (1998). A rigorous MINLP model for the optimal synthesis and operation of utility plants. *Chemical Engineering Research and Design*, 76(3):246–258.

Buoro, D., Casisi, M., De Nardi, A., Pinamonti, P., and Reini, M. (2013). Multicriteria optimization of a distributed energy supply system for an industrial area. *Energy*, 58:128–137.

Carvalho, M., Lozano, M. A., and Serra, L. M. (2012). Multicriteria synthesis of trigeneration systems considering economic and environmental aspects. *Applied Energy*, 91(1):245–254.

Caserta, M. and Voß, S. (2015). An exact algorithm for the reliability redundancy allocation problem. *European Journal of Operational Research*, 244(1):110–116.

Chassein, A. and Goerigk, M. (2016). A bi-criteria approach to robust optimization. *Computers & Operations Research*, 66:181–189.

Chen, S.-J. and Hwang, C.-L. (1992). *Fuzzy Multiple Attribute Decision Making Methods*, pages 289–486. Springer Berlin Heidelberg, Berlin, Heidelberg.

Chicco, G. and Mancarella, P. (2009). Distributed multi-generation: A comprehensive view. *Renewable and Sustainable Energy Reviews*, 13(3):535–551.

Choi, J., Mount, T. D., Thomas, R. J., and Billinton, R. (2006). Probabilistic reliability criterion for planning transmission system expansions. *IEE Proceedings – Generation, Transmission and Distribution*, 153(6):719–727.

Choi, J., Tran, T., El-Keib, A. A., Thomas, R., Oh, H., and Billinton, R. (2005). A method for transmission system expansion planning considering probabilistic reliability criteria. *IEEE Transactions on Power Systems*, 20(3):1606–1615.

Chu, S. and Majumdar, A. (2012). Opportunities and challenges for a sustainable energy future. *nature*, 488:294–303.

Das, I. (1999). A preference ordering among various Pareto optimal alternatives. *Structural and Multidisciplinary Optimization*, 18(1):30–35.

Das, I. and Dennis, J. E. (1998). Normal-boundary intersection: A new method for generating the Pareto surface in nonlinear multicriteria optimization problems. *Siam Journal on Optimization*, 8(3):631–657.

de la Fuente, D., Vega-Rodríguez, M. A., and Pérez, C. J. (2018). Automatic selection of a single solution from the Pareto front to identify key players in social networks. *Knowledge-Based Systems*, 160:228–236.

Deb, K. and Gupta, H. (2004). Introducing robustness in multi-objective optimization. KanGAL Report No. 2004016, Indian Institute of Technology Kanpur.

DeCarolis, J., Daly, H., Dodds, P., Keppo, I., Li, F., McDowall, W., Pye, S., Strachan, N., Trutnevyte, E., Usher, W., Winning, M., Yeh, S., and Zeyringer, M. (2017). Formalizing best practice for energy system optimization modelling. *Applied Energy*, 194:184–198.

Delage, E. and Ye, Y. (2010). Distributionally robust optimization under moment uncertainty with application to data-driven problems. *Operations Research*, 58(3):595–612.

Demirhan, C. D., Tso, W. W., Ogumerem, G. S., and Pistikopoulos, E. N. (2019). Energy systems engineering – A guided tour. *BMC Chemical Engineering*, 1:11.

Domínguez-Muñoz, F., Cejudo-López, J. M., Carrillo-Andrés, A., and Gallardo-Salazar, M. (2011). Selection of typical demand days for CHP optimization. *Energy and Buildings*, 43(11):3036–3043.

Dong, C., Huang, G., Cai, Y., and Liu, Y. (2013). Robust planning of energy management systems with environmental and constraint-conservative considerations under multiple uncertainties. *Energy Conversion and Management*, 65:471–486.

Dong, C., Huang, G. H., Cai, Y. P., and Xu, Y. (2011). An interval-parameter minimax regret programming approach for power management systems planning under uncertainty. *Applied Energy*, 88(8):2835–2845.

Doolittle, E. K., Kerivin, H. L. M., and Wiecek, M. M. (2018). Robust multiobjective optimization with application to internet routing. *Annals of Operations Research*, 271(2):487–525.

Dranichak, G. M. and Wiecek, M. M. (2019). On highly robust efficient solutions to uncertain multiobjective linear programs. *European Journal of Operational Research*, 273(1):20–30.

Duckstein, L. and Opricovic, S. (1980). Multiobjective optimization in river basin development. *Water Resources Research*, 16(1):14–20.

Duran, M. and Grossmann, I. E. (1986). An outer-approximation algorithm for a class of mixed-integer nonlinear programs. *Mathematical Programming*, 36(3):307–339.

Ehrgott, M. (2005). *Multicriteria Optimization*. Springer Berlin Heidelberg, 2nd edition.

Ehrgott, M., Ide, J., and Schöbel, A. (2014). Minmax robustness for multi-objective optimization problems. *European Journal of Operational Research*, 239(1):17–31.

Elsido, C., Bischi, A., Silva, P., and Martelli, E. (2017). Two-stage MINLP algorithm for the optimal synthesis and design of networks of CHP units. *Energy*, 121:403–426.

European Commission (2018). Work programme 2018–2020: 10. Secure, clean and efficient energy. European Commission Decision C(2018)4708.

Fazlollahi, S., Becker, G., and Maréchal, F. (2014a). Multi-objectives, multi-period optimization of district energy systems: III. Distribution networks. *Computers & Chemical Engineering*, 66:82–97.

Fazlollahi, S., Bungener, S. L., Mandel, P., Becker, G., and Maréchal, F. (2014b). Multi-objectives, multi-period optimization of district energy systems: I. Selection of typical operating periods. *Computers & Chemical Engineering*, 65:54–66.

Fazlollahi, S., Mandel, P., Becker, G., and Maréchal, F. (2012). Methods for multi-objective investment and operating optimization of complex energy systems. *Energy*, 45(1):12–22.

Fernandez-Pales, A., Levi, P., and Vass, T. (2019). Industry: Tracking clean energy progress. International Energy Agency.

Fischetti, M. and Monaci, M. (2009). Light robustness. In Ahuja, R. K., Möhring, R. H., and Zaroliagis, C. D., editors, *Robust and Online Large-Scale Optimization*, volume 5868 of *Lecture Notes in Computer Science*, pages 61–84. Springer Berlin Heidelberg.

Fliege, J. and Werner, R. (2014). Robust multiobjective optimization & applications in portfolio optimization. *European Journal of Operational Research*, 234(2):422–433.

Flores, J., Montagna, J. M., and Vecchietti, A. (2015). Investment planning in energy considering economic and environmental objectives. *Computers & Chemical Engineering*, 72:222–232.

Frangopoulos, C. A. (2018). Recent developments and trends in optimization of energy systems. *Energy*, 164:1011–1020.

Frangopoulos, C. A. and Dimopoulos, G. G. (2004). Effect of reliability considerations on the optimal synthesis, design and operation of a cogeneration system. *Energy*, 29(3):309–329.

Frangopoulos, C. A., Spakovsky, M. R. v., and Sciubba, E. (2002). A brief review of methods for the design and synthesis optimization of energy systems. *International Journal of Applied Thermodynamics*, 5(4):151–160.

Fuentes-Cortés, L. F., Santibañez-Aguilar, J. E., and Ponce-Ortega, J. M. (2016). Optimal design of residential cogeneration systems under uncertainty. *Computers & Chemical Engineering*, 88:86–102.

Gabrielli, P., Fürer, F., Mavromatidis, G., and Mazzotti, M. (2019). Robust and optimal design of multi-energy systems with seasonal storage through uncertainty analysis. *Applied Energy*, 238:1192–1210.

Gebreslassie, B. H., Guillén-Gosálbez, G., Jiménez, L., and Boer, D. (2012). Solar assisted absorption cooling cycles for reduction of global warming: A multi-objective optimization approach. *Solar Energy*, 86(7):2083–2094.

Giarola, S., Zamboni, A., and Bezzo, F. (2011). Spatially explicit multi-objective optimisation for design and planning of hybrid first and second generation biorefineries. *Computers & Chemical Engineering*, 35(9):1782–1797.

Glover, F. (1975). Improved linear integer programming formulations of nonlinear integer problems. *Management Science*, 22(4):391–507.

Goberna, M., Jeyakumar, V., Li, G., and Vicente-Pérez, J. (2015). Robust solutions to multi-objective linear programs with uncertain data. *European Journal of Operational Research*, 242(3):730–743.

Goderbauer, S., Bahl, B., Voll, P., Lübbecke, M. E., Bardow, A., and Koster, A. M. (2016). An adaptive discretization MINLP algorithm for optimal synthesis of decentralized energy supply systems. *Computers & Chemical Engineering*, 95:38–48.

Goderbauer, S., Comis, M., and Willamowski, F. J. L. (2019). The synthesis problem of decentralized energy systems is strongly NP-hard. *Computers & Chemical Engineering*, 124:343–349.

Goerigk, M. and Schöbel, A. (2016). Algorithm engineering in robust optimization. In *Algorithm Engineering: Selected Results and Surveys*, volume 9220, pages 245–279, Cham. Springer International Publishing.

Gong, J. and You, F. (2018). Resilient design and operations of process systems: Nonlinear adaptive robust optimization model and algorithm for resilience analysis and enhancement. *Computers & Chemical Engineering*, 116:231–252.

Grossmann, I. E., Apap, R. M., Calfa, B. A., García-Herreros, P., and Zhang, Q. (2016). Recent advances in mathematical programming techniques for the optimization of process systems under uncertainty. *Computers & Chemical Engineering*, 91:3–14.

Grossmann, I. E. and Guillén-Gosálbez, G. (2010). Scope for the application of mathematical programming techniques in the synthesis and planning of sustainable processes. *Computers & Chemical Engineering*, 34(9):1365–1376.

Guillén-Gosálbez, G. (2011). A novel MILP-based objective reduction method for multi-objective optimization: Application to environmental problems. *Computers & Chemical Engineering*, 35(8):1469–1477.

Guo, L., Liu, W., Cai, J., Hong, B., and Wang, C. (2013). A two-stage optimal planning and design method for combined cooling, heat and power microgrid system. *Energy Conversion and Management*, 74:433–445.

Guzman, Y. A., Matthews, L. R., and Floudas, C. A. (2016). New a priori and a posteriori probabilistic bounds for robust counterpart optimization: I. Unknown probability distributions. *Computers & Chemical Engineering*, 84:568–598.

Guzman, Y. A., Matthews, L. R., and Floudas, C. A. (2017). New a priori and a posteriori probabilistic bounds for robust counterpart optimization: II. A priori bounds for known symmetric and asymmetric probability distributions. *Computers & Chemical Engineering*, 101:279–311.

Hedman, K. W., Ferris, M. C., O'Neill, R. P., Fisher, E. B., and Oren, S. S. (2010). Co-optimization of generation unit commitment and transmission switching with $N-1$ reliability. *IEEE Transactions on Power Systems*, 25(2):1052–1063.

Hennen, M., Postels, S., Voll, P., Lampe, M., and Bardow, A. (2017). Multi-objective synthesis of energy systems: Efficient identification of design trade-offs. *Computers & Chemical Engineering*, 97:283–293.

Hollermann, D. E., Goerigk, M., Hoffrogge, D. F., Hennen, M., and Bardow, A. (2019a). Flexible here-and-now decisions for two-stage multi-objective optimization: Method and application to energy system design selection. *arXiv e-prints*, arXiv:1906.08621.

Hollermann, D. E., Hoffrogge, D. F., Mayer, F., Hennen, M., and Bardow, A. (2019b). Optimal $(n-1)$-reliable design of distributed energy supply systems. *Computers & Chemical Engineering*, 121:317–326.

Hong, S., Cheng, H., and Zeng, P. (2017). $N-K$ constrained composite generation and transmission expansion planning with interval load. *IEEE Access*, 5:2779–2789.

Huijbregts, M. A. J., Steinmann, Z. J. N., Elshout, P. M. F., Stam, G., Verones, F., Vieira, M. D. M., Hollander, A., Zijp, M., and van Zelm, R. (2016). ReCiPe 2016: A harmonized life cycle impact assessment method at midpoint and endpoint level report I: Characterization.

Hwang, C.-L. and Yoon, K. (1981). *Methods for Multiple Attribute Decision Making*, pages 58–191. Springer Berlin Heidelberg, Berlin, Heidelberg.

IBM Corporation (2015). IBM ILOG CPLEX Optimization Studio, Version 12.6. User Guide.

Ide, J. and Schöbel, A. (2016). Robustness for uncertain multi-objective optimization: A survey and analysis of different concepts. *OR Spectrum*, 38(1):235–271.

Ihlemann, M. (2017). *$(n-1)$-reliable structure optimization of decentralized energy systems considering start-up and shut-down processes.* Master thesis, RWTH Aachen University.

Jahromi, A. E. and Feizabadi, M. (2017). Optimization of multi-objective redundancy allocation problem with non-homogeneous components. *Computers & Industrial Engineering*, 108:111–123.

Jennings, M., Fisk, D., and Shah, N. (2014). Modelling and optimization of retrofitting residential energy systems at the urban scale. *Energy*, 64:220–233.

Ji, L., Niu, D. X., and Huang, G. H. (2014). An inexact two-stage stochastic robust programming for residential micro-grid management-based on random demand. *Energy*, 67:186–199.

Jing, R., Wang, M., Zhang, Z., Liu, J., Liang, H., Meng, C., Shah, N., Li, N., and Zhao, Y. (2019a). Comparative study of posteriori decision-making methods when designing building integrated energy systems with multi-objectives. *Energy and Buildings*, 194:123–139.

Jing, R., Wang, M., Zhang, Z., Wang, X., Li, N., Shah, N., and Zhao, Y. (2019b). Distributed or centralized? Designing district-level urban energy systems by a hierarchical approach considering demand uncertainties. *Applied Energy*, 252:113424.

Karimi, H. and Jadid, S. (2019). Optimal microgrid operation scheduling by a novel hybrid multi-objective and multi-attribute decision-making framework. *Energy*, 186:115–912.

Karmellos, M., Georgiou, P., and Mavrotas, G. (2019). A comparison of methods for the optimal design of distributed energy systems under uncertainty. *Energy*, 178:318–333.

Kaundinya, D. P., Balachandra, P., and Ravindranath, N. H. (2009). Grid-connected versus stand-alone energy systems for decentralized power – A review of literature. *Renewable and Sustainable Energy Reviews*, 13(8):2041–2050.

Kloepffer, W. (2008). Life cycle sustainability assessment of products. *International Journal of Life Cycle Assessment*, 13(2):89–95.

Kohler, S., Agricola, A.-C., and Seidl, H. (2010). dena-Netzstudie II. Integration erneuerbarer Energien in die deutsche Stromversorgung im Zeitraum 2015–2020 mit Ausblick 2025. Deutsche Energie-Agentur GmbH (dena).

Kotzur, L., Markewitz, P., Robinius, M., and Stolten, D. (2018). Impact of different time series aggregation methods on optimal energy system design. *Renewable Energy*, 117(C):474–487.

Kuhn, K., Raith, A., Schmidt, M., and Schöbel, A. (2016). Bi-objective robust optimisation. *European Journal of Operational Research*, 252(2):418–431.

Kuroiwa, D. and Lee, G. M. (2012). On robust multiobjective optimization. *Vietnam Journal of Mathematics*, 40(2&3):305–317.

Lemos, L. P., Lima, E. L., and Pinto, J. C. (2018). New decision making criterion for multiobjective optimization problems. *Industrial & Engineering Chemistry Research*, 57(3):1014–1025.

Leontief, W. W. (1936). Quantitative input and output relations in the economic systems of the united states. *The Review of Economics and Statistics*, 18(3):105–125.

Li, Z., Ding, R., and Floudas, C. A. (2011). A comparative theoretical and computational study on robust counterpart optimization: I. Robust linear optimization and robust mixed integer linear optimization. *Industrial & Engineering Chemistry Research*, 50(18):10567–10603.

Li, Z., Liao, H., and Coit, D. W. (2009). A two-stage approach for multi-objective decision making with applications to system reliability optimization. *Reliability Engineering & System Safety*, 94(10):1585–1592.

Li, Z., Wu, W., Shahidehpour, M., Wang, J., and Zhang, B. (2016). Combined heat and power dispatch considering pipeline energy storage of district heating network. *IEEE Transactions on Sustainable Energy*, 7:12–22.

Lin, F., Leyffer, S., and Munson, T. (2016). A two-level approach to large mixed-integer programs with application to cogeneration in energy-efficient buildings. *Computational Optimization and Applications*, 65(1):1–46.

Majewski, D. E., Lampe, M., Voll, P., and Bardow, A. (2017a). TRusT: A two-stage robustness trade-off approach for the design of decentralized energy supply systems. *Energy*, 118:590–599.

Majewski, D. E., Wirtz, M., Lampe, M., and Bardow, A. (2017b). Robust multi-objective optimization for sustainable design of distributed energy supply systems. *Computers & Chemical Engineering*, 102:26–39.

Mancarella, P. (2014). MES (multi-energy systems): An overview of concepts and evaluation models. *Energy*, 65:1–17.

Månsson, A., Johansson, B., and Nilsson, L. J. (2014). Assessing energy security: An overview of commonly used methodologies. *Energy*, 73:1–14.

Mattson, C. A. and Messac, A. (2003). Concept selection using s-Pareto frontiers. *AIAA Journal*, 41(6):1190–1198.

Mavrotas, G. (2009). Effective implementation of the epsilon-constraint method in multiobjective mathematical programming problems. *Applied Mathematics and Computation*, 213(2):455–465.

Mayer, F. (2018). *Two-stage robust and $(n-1)$-reliable optimization of distributed energy supply systems*. Master thesis, RWTH Aachen University.

McCarl, B. A. (2014). McCarl GAMS User Guide, Version 24.3.

McCarl, B. A. and Rosenthal, R. E. (2016). McCarl GAMS User Guide, Version 24.7.

McKay, M. D., Beckman, R. J., and Conover, W. J. (2000). A comparison of three methods for selecting values of input variables in the analysis of output from a computer code. *Technometrics*, 42(1):55–61.

Miettinen, K. (2008). *Introduction to multiobjective optimization: Noninteractive approaches*, pages 1–26. Springer Berlin Heidelberg, Berlin, Heidelberg.

Minciuc, E., Le Corre, O., Athanasovici, V., Tazerout, M., and Bitir, I. (2003). Thermodynamic analysis of tri-generation with absorption chilling machine. *Applied Thermal Engineering*, 23(11):1391–1405.

Miryousefi Aval, S. M., Ahadi, A., and Hayati, H. (2015). Adequacy assessment of power systems incorporating building cooling, heating and power plants. *Energy and Buildings*, 105:236–246.

Mínguez, R. and García-Bertrand, R. (2016). Robust transmission network expansion planning in energy systems: Improving computational performance. *European Journal of Operational Research*, 248(1).21–32.

Moreira, A., Street, A., and Arroyo, J. M. (2015). An adjustable robust optimization approach for contingency-constrained transmission expansion planning. *IEEE Transactions on Power Systems*, 30(4):2013–2022.

Moret, S. (2017). *Strategic energy planning under uncertainty*. Phd thesis, École Polytechnique Fédérale de Lausanne.

Moret, S., Babonneau, F., Bierlaire, M., and Maréchal, F. (2019). Decision support for strategic energy planning: A robust optimization framework. *European Journal of Operational Research.*

Moret, S., Bierlaire, M., and Maréchal, F. (2014). Robust optimization for strategic energy planning. Technical Report TRANSP-OR 141115, Transport and Mobility Laboratory, Ecole Polytechnique Fédérale de Lausanne.

Mota, B., Gomes, M. I., Carvalho, A., and Barbosa-Povoa, A. P. (2015). Towards supply chain sustainability: Economic, environmental and social design and planning. *Journal of Cleaner Production*, 105:14–27.

Ning, C. and You, F. (2017). A data-driven multistage adaptive robust optimization framework for planning and scheduling under uncertainty. *AIChE Journal*, 63(10):4343–4369.

Ning, C. and You, F. (2018). Adaptive robust optimization with minimax regret criterion: Multiobjective optimization framework and computational algorithm for planning and scheduling under uncertainty. *Computers & Chemical Engineering*, 108:425–447.

Nitsch, J., Pregger, T., Naegler, T., Heide, D., de Tena, D. L., Trieb, F., and et al. (2012). Langfristszenarien und Strategien für den Ausbau der erneuerbaren Energien in Deutschland bei Berücksichtigung der Entwicklung in Europa und global. Deutsches Zentrum für Luft- und Raumfahrt (DLR) (Schlussbericht BMU - FKZ 03MAP146).

Opricovic, S. and Tzeng, G.-H. (2004). Compromise solution by MCDM methods: A comparative analysis of VIKOR and TOPSIS. *European Journal of Operational Research*, 156(2):445–455.

Padhye, N. and Deb, K. (2011). *Multi-objective Optimisation and Multi-criteria Decision Making for FDM Using Evolutionary Approaches*, pages 219–247. Springer London, London.

Pepermans, G., Driesen, J., Haeseldonckx, D., Belmans, R., and D'haeseleer, W. (2005). Distributed generation: Definition, benefits and issues. *Energy Policy*, 33(6):787–798.

Pohekar, S. and Ramachandran, M. (2004). Application of multi-criteria decision making to sustainable energy planning – A review. *Renewable and Sustainable Energy Reviews*, 8(4):365–381.

Popovic, T., Barbosa-Póvoa, A., Kraslawski, A., and Carvalho, A. (2018). Quantitative indicators for social sustainability assessment of supply chains. *Journal of Cleaner Production*, 180:748–768.

Pérez, C. J., Vega-Rodríguez, M. A., Reder, K., and Flörke, M. (2017). A multi-objective artificial bee colony-based optimization approach to design water quality monitoring networks in river basins. *Journal of Cleaner Production*, 166:579–589.

Pulsipher, J. L., Rios, D., and Zavala, V. M. (2019). A computational framework for quantifying and analyzing system flexibility. *Computers & Chemical Engineering*, 126:342–355.

Quintana, D., Denysiuk, R., Garcia-Rodriguez, S., and Gaspar-Cunha, A. (2017). Portfolio implementation risk management using evolutionary multiobjective optimization. *Applied Science*, 7(10):1079.

Rachmawati, L. and Srinivasan, D. (2009). Multiobjective evolutionary algorithm with controllable focus on the knees of the Pareto front. *IEEE Transactions on Evolutionary Computation*, 13(4):810–824.

Ramos, T. R. P., Gomes, M. I., and Barbosa-Póvoa, A. P. (2014). Planning a sustainable reverse logistics system: Balancing costs with environmental and social concerns. *Omega*, 48:60–74.

Ruiz, C. and Conejo, A. (2015). Robust transmission expansion planning. *European Journal of Operational Research*, 242(2):390–401.

Ruiz, P. A., Philbrick, C. R., Zak, E., Cheung, K. W., and Sauer, P. W. (2009). Uncertainty management in the unit commitment problem. *IEEE Transactions on Power Systems*, 24(2):642–651.

Salcedo, R., Antipova, E., Boer, D., Jiménez, L., and Guillén-Gosálbez, G. (2012). Multi-objective optimization of solar Rankine cycles coupled with reverse osmosis desalination considering economic and life cycle environmental concerns. *Desalination*, 286:358–371.

Sass, S., Faulwasser, T., Hollermann, D. E., Kappatou, C. D., Sauer, D., Schütz, T., Shu, D. Y., Bardow, A., Gröll, L., Hagenmeyer, V., Müller, D., and Mitsos, A. (2020). Model compendium, data, and optimization benchmarks for sector-coupled energy systems. *Computers & Chemical Engineering*, 135:106760.

Schöbel, A. (2014). Generalized light robustness and the trade-off between robustness and nominal quality. *Mathematical Methods of Operations Research*, 80(2):161–191.

Schütz, T., Schiffer, L., Harb, H., Fuchs, M., and Müller, D. (2017). Optimal design of energy conversion units and envelopes for residential building retrofits using a comprehensive MILP model. *Applied Energy*, 185:1–15.

Shang, Z. and Kokossis, A. (2005). A systematic approach to the synthesis and design of flexible site utility systems. *Chemical Engineering Science*, 60(16):4431–4451.

Shi, H. and You, F. (2016). A computational framework and solution algorithms for two-stage adaptive robust scheduling of batch manufacturing processes under uncertainty. *AIChE Journal*, 62:687–703.

Soyster, A. L. (1973). Technical note – Convex programming with set-inclusive constraints and applications to inexact linear programming. *Operations Research*, 21(5):1154–1157.

Srinivasan, V. and Shocker, A. D. (1973). Linear programming techniques for multidimensional analysis of preferences. *Psychometrika*, 38(3):337–369.

Street, A., Moreira, A., and Arroyo, J. M. (2014). Energy and reserve scheduling under a joint generation and transmission security criterion: An adjustable robust optimization approach. *IEEE Transactions on Power Systems*, 29(1):3–14.

Street, A., Oliveira, F., and Arroyo, J. M. (2011). Contingency-constrained unit commitment with $n - K$ security criterion: A robust optimization approach. *IEEE Transactions on Power Systems*, 26(3):1581–1590.

Sun, G., Zhang, H., Fang, J., Li, G., and Li, Q. (2018). A new multi-objective discrete robust optimization algorithm for engineering design. *Applied Mathematical Modelling*, 53:602–621.

Sun, L., Gai, L., and Smith, R. (2017). Site utility system optimization with operation adjustment under uncertainty. *Applied Energy*, 186(3):450–456.

Sun, L. and Liu, C. (2015). Reliable and flexible steam and power system design. *Applied Thermal Engineering*, 79:184–191.

Sy, C. L., Aviso, K. B., Ubando, A. T., and Tan, R. R. (2016). Target-oriented robust optimization of polygeneration systems under uncertainty. *Energy*, 116(2):1334–1347.

Taboada, H. A., Baheranwala, F., Coit, D. W., and Wattanapongsakorn, N. (2007). Practical solutions for multi-objective optimization: An application to system reliability design problems. *Reliability Engineering & System Safety*, 92(3):314–322.

Thiele, A., Terry, T., and Epelman, M. (2010). Robust linear optimization with recourse. Technical report, Lehigh University.

thinkstep (2016). GaBi Software-System and databases for life cycle engineering. thinkstep AG.

Tock, L. and Maréchal, F. (2015). Decision support for ranking Pareto optimal process designs under uncertain market conditions. *Computers & Chemical Engineering*, 83:165–175.

Ünal, A. N., Ercan, S., and Kayakutlu, G. (2015). Optimisation studies on tri-generation: A review. *International Journal of Energy Research*, 39(10):1311–1334.

United Nations (2015). World urbanization prospects: The 2014 revision. Technical report, Department of Economic and Social Affairs, Population Division. Technical Report ST/ESA/SER.A/366.

Vallerio, M., Hufkens, J., van Impe, J., and Logist, F. (2015). An interactive decision-support system for multi-objective optimization of nonlinear dynamic processes with uncertainty. *Expert Systems with Applications*, 42(21):7710–7731.

Vallerio, M., Telen, D., Cabianca, L., Manenti, F., van Impe, J., and Logist, F. (2016). Robust multi-objective dynamic optimization of chemical processes using the sigma point method. *Chemical Engineering Science*, 140:201–216.

Voll, P. (2013). *Automated Optimization-Based Synthesis of Distributed Energy and Supply Systems.* PhD thesis, RWTH Aachen University.

Voll, P., Klaffke, C., Hennen, M., and Bardow, A. (2013). Automated superstructure-based synthesis and optimization of distributed energy supply systems. *Energy*, 50:374–388.

Wakui, T., Hashiguchi, M., Sawada, K., and Yokoyama, R. (2019). Two-stage design optimization based on artificial immune system and mixed-integer linear programming for energy supply networks. *Energy*, 170:1228–1248.

Wang, Q., Watson, J. P., and Guan, Y. (2013). Two-stage robust optimization for $N-k$ contingency constrained unit commitment. *IEEE Transactions on Power Systems*, 28(3):2366–2375.

Wang, Y., Wang, Y., Huang, Y., Li, F., Zeng, M., Li, J., Wang, X., and Zhang, F. (2019). Planning and operation method of the regional integrated energy system considering economy and environment. *Energy*, 171:731–750.

Weber, C. I. (2008). *Multi-objective design and optimization of district energy systems including polygeneration energy conversion technologies.* PhD thesis, École Polytechnique Fédérale de Lausanne, Lausanne.

Wiest, P., Rudion, K., and Probst, A. (2018). Efficient integration of $(n-1)$-security into probabilistic network expansion planning. *International Journal of Electrical Power & Energy Systems*, 94:151–159.

Yaman, H. and. Karaşan, O. and Pinar, M. (2001). The robust spanning tree problem with interval data. *Operations Research Letters*, 29(1):31–40.

Yanıkoğlu, İ., Gorissen, B. L., and den Hertog, D. (2019). A survey of adjustable robust optimization. *European Journal of Operational Research*, 277(3):799–813.

Ye, Y., Grossmann, I. E., and Pinto, J. M. (2017). Mixed-integer nonlinear programming models for optimal design of reliable chemical plants. *Computers & Chemical Engineering*, 116:3–16.

Ye, Y., Grossmann, I. E., Pinto, J. M., and Ramaswamy, S. (2019). Modeling for reliability optimization of system design and maintenance based on Markov chain theory. *Computers & Chemical Engineering*, 124:381–404.

Yokoyama, R., Fujiwara, K., Ohkura, M., and Wakui, T. (2014). A revised method for robust optimal design of energy supply systems based on mini-max regret criterion. *Energy Conversion and Management*, 84:196–208.

Yokoyama, R., Shinano, Y., Wakayama, Y., and Wakui, T. (2019). Model reduction by time aggregation for optimal design of energy supply systems by an MILP hierarchical branch and bound method. *Energy*, 181:782–792.

Yokoyama, R., Tokunaga, A., and Wakui, T. (2018). Robust optimal design of energy supply systems under uncertain energy demands based on a mixed-integer linear model. *Energy*, 153:159–169.

Zelany, M. (1974). A concept of compromise solutions and the method of the displaced ideal. *Computers & Operations Research*, 1(3–4):479–496.

Zhang, Q., Grossmann, I. E., and Lima, R. M. (2016). On the relation between flexibility analysis and robust optimization for linear systems. *AIChE Journal*, 62(9):3109–3123.

Zhao, L. and You, F. (2019). A data-approach for industrial utility systems optimization under uncertainty. *Energy*, 182:559–569.

Ziher, D. and Poredos, A. (2006). Economics of a trigeneration system in a hospital. *Applied Thermal Engineering*, 26(7):680–687.

Zio, E. and Bazzo, R. (2011). A clustering procedure for reducing the number of representative solutions in the Pareto front of multiobjective optimization problems. *European Journal of Operational Research*, 210(3):624–634.

Zitzler, E., Thiele, L., Laumanns, M., Fonseca, C. M., and da Fonseca, V. G. (2003). Performance assessment of multiobjective optimizers: An analysis and review. *IEEE Transactions on Evolutionary Computation*, 7(2):117–132.

Aachener Beiträge zur Technischen Thermodynamik

ABTT 1
Philip Voll
Automated Optimization-Based Synthesis of Distributed Energy Supply Systems
1. Auflage 2014
ISBN 978-3-86130-474-6

ABTT 2
Johannes Jung
Comparative Life Cycle Assessment of Industrial Multi-Product Processes
1. Auflage 2014
ISBN 978-3-86130-471-5

ABTT 3
Franz Lanzerath
Modellgestützte Entwicklung von Adsorptionswärmepumpen
1. Auflage 2014
ISBN 978-3-86130-472-2

ABTT 4
Thorsten Brands
Einfluss der Gemischzusammensetzung auf die Verbrennung im Diesel- und GCAI-Motor
1. Auflage 2014
ISBN 978-3-95886-006-3

ABTT 5
Dominique Dechambre
Efficient Measurement of Liquid-Liquid Equilibria using Automation and Optimal Experimental Design
1. Auflage 2016
ISBN 978-395886-077-3

ABTT 6
Niklas von der Aßen
From Life-Cycle Assesement towards life-Cycle Design of Carbon Dioxide Capture and Utilization
1. Auflage 2016
ISBN 978-3-95886-080-3

ABTT 7
Matthias Lampe
Integrated Process and Organic Rankine Cycle Working Fluid Design in the Continuous-Molecular Targeting Framework
1. Auflage 2016
ISBN 978-3-95886-086-5

ABTT 8
Thomas Hülser
Optische Untersuchung der Zündvorgänge und deren Auswirkung auf die Verbrennung in PKW-Motoren
1. Auflage 2016
ISBN 978-3-95886-090-2

Aachener Beiträge zur Technischen Thermodynamik

ABTT 9
Malte Döntgen
Reaction Models from Reactive Molecular Dynamics and High-Level Kinetics Predictions
1. Auflage 2016
ISBN 978-3-95886-156-5

ABTT 10
Heike Schreiber
Experiments and Validated Models for Adsorption Thermal Energy Storage in Industrial and Residential Application
1. Auflage 2017
ISBN 978-3-95886-178-7

ABTT 11
André Dirk Sternberg
System-Wide Perspective for Life Cycle Assesment of CO_2-based C1-Chemicals
1. Auflage 2017
ISBN 978-3-95886-193-0

ABTT 12
Uwe Bau
From Dynamic Simulation to Optimal Design and Control of Adsorption Energy Systems
1. Auflage 2018
ISBN 978-3-95886-216-6

ABTT 13
Christian Jens
Modellbasiertes Design von Produkt, Lösungsmittel und Prozess für die Ameisensäure-synthese aus CO_2 und H_2
1. Auflage 2018
ISBN 978-3-95886-231-9

ABTT 14
Jan David Scheffczyk
Integrated Computer-Aided Design of Molecules and Processes using COSMO-RS
1. Auflage 2018
ISBN 978-3-95886-236-4

ABTT 15
Björn Bahl
Optimization-Based Synthesis of Large-Scale Energy Systems by Time-Series Aggregation
1. Auflage 2018
ISBN 978-3-95886-240-1

Aachener Beiträge zur Technischen Thermodynamik

ABTT 16
Bastian Liebergesell
A Milliliter-Scale Setup for the Efficient Characterization of Multicomponent Vapor-Liquid Equilibria Using Raman Spectroscopy
1. Auflage 2018
ISBN 978-3-95886-247-0

ABTT 17
Stefan Wilhelm Graf
A Design Approach for Adsorption Energy Systems Integrating Dynamic Modeling with Small-Scale Experiments
1. Auflage 2018
ISBN 978-3-95886-258-6

ABTT 18
Sebastian Kaminski
Quantum-Mechanics-Based Prediction of SAFT Parameters for Non-Associating and Associating Molecules Containing Carbon, Hydrogen, Oxygen and Nitrogen
1. Auflage 2019
ISBN 978-3-95886-270-8

ABTT 19
Maike Renate Hennen
Decision Support for the Synthesis of Energy Systems by Analysis of the Near-Optimal Solution Space
1. Auflage 2019
ISBN 978-3-95886-277-7

ABTT 20
Peyman Yamin
COSMO-RS-Based Methods for Improved Modelling of Complex Chemical Systems
1. Auflage 2019
ISBN 978-3-95886-288-3

ABTT 21
Meltem Erdogan
Assessement of Adsorbents for Drying by Experiments and Dynamic Simulations
1. Auflage 2019
ISBN 978-3-95886-303-3

ABTT 22
Christian Schulz
SRS/LIF-Messungen zur Charakterisierung rußarmer dieselähnlicher Flammen von alternativen Kraftstoffen und n-Heptan
1. Auflage 2019
ISBN 978-3-91886-310-1

Aachener Beiträge zur Technischen Thermodynamik

ABTT 23
Peter Beumers
Physically-Based Models for the Analysis of Raman Spectra
1. Auflage 2019
ISBN 978-3-95886-319-4

ABTT 24
Arne Kätelhön
Technology Choice Model for Consequential Life Cycle Assessment
1. Auflage 2019
ISBN 978-3-95886-324-8

ABTT 25
Christine Peters
Measurement of Multicomponent Diffusion in Liquids Using Raman Microspectroscopy and Microfluidics
1. Auflage 2020
ISBN 978-3-95886-337-8

ABTT 26
Dinah Elena Hollermann
Reliable and Robust Optimal Design of Sustainable Energy Systems
1. Auflage 2020
ISBN 978-3-95886-346-0